SAXON

Algebra 1, Geometry, Algebra 2

College Entrance Exam Practice

An Imprint of HMH
Supplemental Publishers

www.SaxonPublishers.com
1-800-531-5015

ISBN 13: 978-1-6027-7495-7

ISBN 10: 1-6027-7495-1

Printed in the United States of America

1 2 3 4 5 6 7 8 170 15 14 13 12 11 10 09 08

Contents

Name ____________________ Date __________ Class __________

PSAT Practice Test

1. YOUR NAME: ____________________
(Print) Last First M.I.

SIGNATURE: ____________ **DATE:** ________

HOME ADDRESS: ____________________
(Print) Number and Street

____________________ **E-MAIL:** ____________
City State Zip

PHONE NO.: ________ **SCHOOL:** ________ **CLASS OF:** ________
(Print)

Completely darken bubbles with a No. 2 pencil. If you make a mistake, be sure to erase mark completely. Erase all stray marks.

IMPORTANT: Please fill in these boxes exactly as shown on the back cover of your text book.

2. TEST FORM

3. TEST CODE

0	0	0	0
1	1	1	1
2	2	2	2
3	3	3	3
4	4	4	4
5	5	5	5
6	6	6	6
7	7	7	7
8	8	8	8
9	9	9	9

4. PHONE NUMBER

0	0	0	0	0	0	0
1	1	1	1	1	1	1
2	2	2	2	2	2	2
3	3	3	3	3	3	3
4	4	4	4	4	4	4
5	5	5	5	5	5	5
6	6	6	6	6	6	6
7	7	7	7	7	7	7
8	8	8	8	8	8	8
9	9	9	9	9	9	9

5. YOUR NAME

First 4 letters of last name				FIRST INT	LAST INT
A	A	A	A	A	A
B	B	B	B	B	B
C	C	C	C	C	C
D	D	D	D	D	D
E	E	E	E	E	E
F	F	F	F	F	F
G	G	G	G	G	G
H	H	H	H	H	H
I	I	I	I	I	I
J	J	J	J	J	J
K	K	K	K	K	K
L	L	L	L	L	L
M	M	M	M	M	M
N	N	N	N	N	N
O	O	O	O	O	O
P	P	P	P	P	P
Q	Q	Q	Q	Q	Q
R	R	R	R	R	R
S	S	S	S	S	S
T	T	T	T	T	T
U	U	U	U	U	U
V	V	V	V	V	V
W	W	W	W	W	W
X	X	X	X	X	X
Y	Y	Y	Y	Y	Y
Z	Z	Z	Z	Z	Z

6. DATE OF BIRTH

MONTH	DAY		YEAR	
JAN				
FEB				
MAR	0	0	0	0
APR	1	1	1	1
MAY	2	2	2	2
JUN	3	3	3	3
JUL		4	4	4
AUG		5	5	5
SEP		6	6	6
OCT		7	7	7
NOV		8	8	8
DEC		9	9	9

7. SEX

- MALE
- FEMALE

8. OTHER

1	A	B	C	D	E
2	A	B	C	D	E
3	A	B	C	D	E

1

MATHEMATICS

1	A B C D E	6	A B C D E	11	A B C D E	16	A B C D E
2	A B C D E	7	A B C D E	12	A B C D E	17	A B C D E
3	A B C D E	8	A B C D E	13	A B C D E	18	A B C D E
4	A B C D E	9	A B C D E	14	A B C D E	19	A B C D E
5	A B C D E	10	A B C D E	15	A B C D E	20	A B C D E

Name ______________________ Date __________ Class __________

PSAT Practice Test

Use a No. 2 pencil only. Be sure each mark is dark and completely fills the intended oval. Completely erase any errors o stray marks.

2 MATHEMATICS

1 (A)(B)(C)(D)(E)
2 (A)(B)(C)(D)(E)
3 (A)(B)(C)(D)(E)
4 (A)(B)(C)(D)(E)
5 (A)(B)(C)(D)(E)
6 (A)(B)(C)(D)(E)
7 (A)(B)(C)(D)(E)
8 (A)(B)(C)(D)(E)

ONLY ANSWERS ENTERED IN THE OVALS IN EACH GRID AREA WILL BE SCORED.
YOU WILL NOT RECEIVE CREDIT FOR ANYTHING WRITTEN IN THE BOXES ABOVE THE OVALS.

9 10 11 12 13

14 15 16 17 18

Name ______________________ Date ____________ Class ____________

PSAT Practice Test 1 Section 1

Time—25 minutes, 20 Questions

Directions: In this section, solve each problem using any available space on the page for scratch work. Then decide which is best of the choices given and fill in the corresponding oval on the answer sheet.

Notes:

1. The use of a calculator is permitted. All numbers used are real numbers.
2. Figures that accompany problems in this test are intended to provide information useful in solving the problems. They are drawn as accurately as possible EXCEPT when it is stated in a specific problem that the figure is not drawn to scale. All figures lie in a plane unless otherwise indicated.

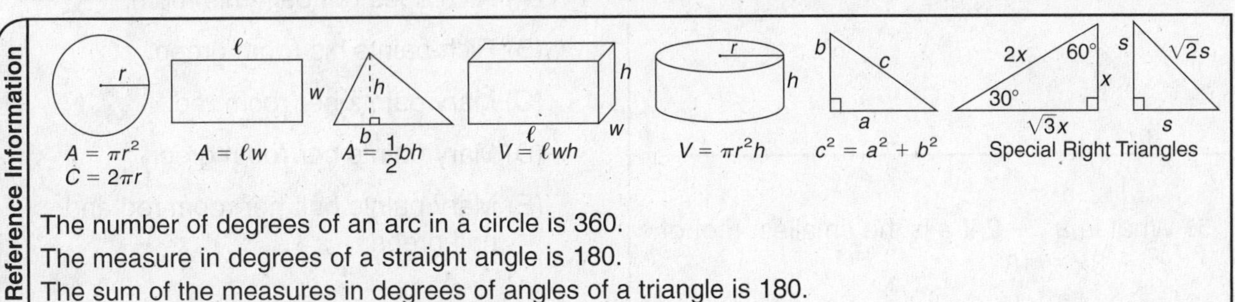

The number of degrees of an arc in a circle is 360.
The measure in degrees of a straight angle is 180.
The sum of the measures in degrees of angles of a triangle is 180.

1. Joe was paid by his mother to clean out his family's garage. He gave 30 percent of his pay to his sister Samantha for boxing up all the old books. Samantha gave 20 percent of her pay to her twin brother Andrew for stacking the boxes of books. If Samantha gave Andrew \$9.00, how much was Joe's pay?

(A) \$45

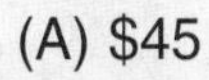

(B) \$75

(C) \$100

(D) \$125

(E) \$150

2. When a number n is subtracted from 18 and the difference is divided by n, the result is 2. What is the value of n?

(A) 4

(B) 6

(C) 8

(D) 10

(E) 14

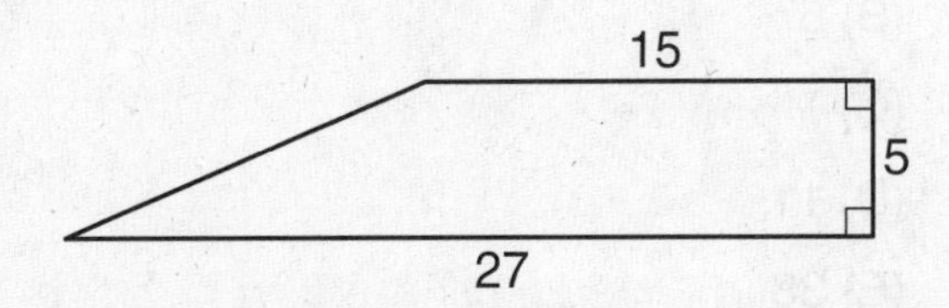

3. What is the perimeter of the trapezoid shown above?

(A) 47

(B) 52

(C) 57

(D) 60

(E) 64

GO ON

PSAT Practice Test 1 Section 1 *continued*

4. If n is a real number, then $2n$ would never be part of which number set?

(A) a real number

(B) an irrational number

(C) a rational number

(D) an imaginary number

(E) an integer

5. What is $a^2 + 2$ if a is the smallest root of $x^2 - 3x - 18$?

(A) -3

(B) 6

(C) 7

(D) 11

(E) 38

6. The roots of the equation $x^2 - 8x + 15 = 0$ are the lengths of the legs of a right triangle. What is the length of the hypotenuse?

(A) $\sqrt{34}$

(B) $\sqrt{35}$

(C) 6

(D) $\sqrt{37}$

(E) $\sqrt{38}$

7. Mary wants to paint her room either red or green; Ben wants to paint his room green; Rich wants to paint his room red. There is just enough red paint to paint only one room and just enough green paint to paint only one room. If Ben does not paint his room, but Mary and Rich both get what they want, then which of the following MUST be true?

(A) Rich does not paint his room.

(B) Rich paints his room green.

(C) Mary paints her room red.

(D) Mary paints her room green.

(E) Mary paints half her room red and half green.

2, 0, 8, -1, 0, 5, -1, -1

8. Using the data above, which of the following is written from the least value to the greatest?

(A) mean, median, mode

(B) mean, mode, median

(C) median, mean, mode

(D) median, mode, mean

(E) mode, median, mean

GO ON

Name ______________________ Date __________ Class __________

PSAT Practice Test 1 Section 1 *continued*

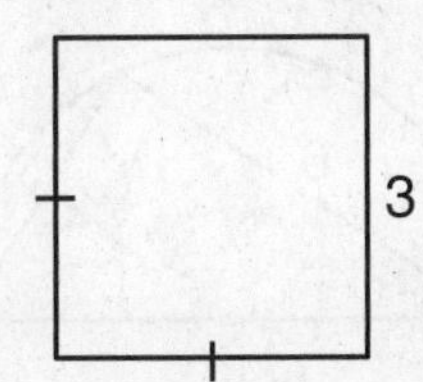

9. The radius of a circle is the same as the perimeter of the square shown above. What is the area of the circle?

(A) 3π

(B) 9π

(C) 12π

(D) 24π

(E) 144π

10. What is an equation of the line that passes through the origin and the intersection of the lines $y = 3x + 4$ and $x = -1$?

(A) $y = -x$

(B) $y = x$

(C) $y = -x + 1$

(D) $y = -x - 1$

(E) $y = x - 1$

11. Which of the following has the smallest value?

(A) $\sqrt{\frac{1}{9}}$

(B) 0.1

(C) $\left(\frac{1}{10}\right)^2$

(D) $\left(\frac{1}{2}\right)^3$

(E) $\frac{7}{21}$

x	$f(x)$
3	4
5	7
7	11
9	16
11	21

12. If $f(a) = 11$, then use the table above to find $f(a - 2)$.

(A) 4

(B) 7

(C) 9

(D) 16

(E) 21

GO ON

Name ______________________ Date ____________ Class ____________

13. The graph of $f(x + 4)$ would be identical to the graph of $f(x)$ except that it would be shifted 4 units

(A) to the right.

(B) to the left.

(C) up.

(D) down.

(E) not enough information

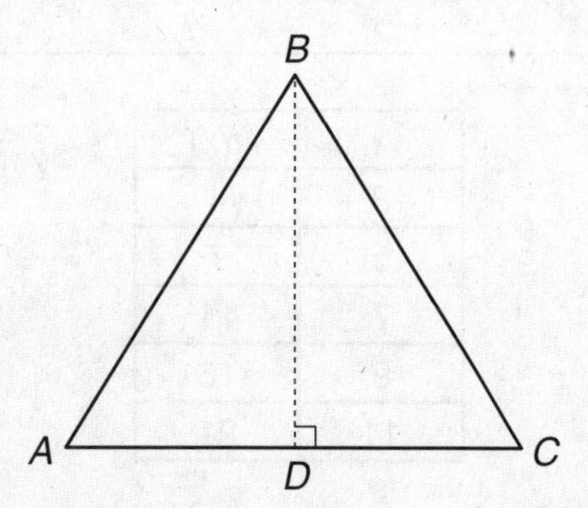

14. The slope of $\overline{AB}$ is $\frac{2}{5}$, the slope of $\overline{BC}$ is $-\frac{2}{5}$, and the length of $\overline{AC}$ is 20. What is the length of $\overline{BD}$?

(A) 2

(B) 4

(C) 5

(D) 8

(E) 10

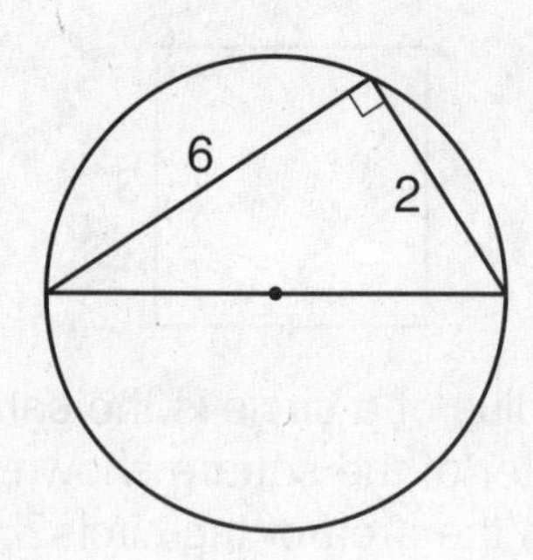

15. What is the radius of the circle shown above?

(A) $\sqrt{2}$

(B) 2

(C) $2\sqrt{2}$

(D) $\sqrt{10}$

(E) $6\sqrt{10}$

$$2x + 3 = 6x - 7$$

16. If a is the solution to the equation above, what is the value of $a^2 + a$?

(A) $4\frac{3}{8}$

(B) 5

(C) $7\frac{1}{2}$

(D) $8\frac{3}{4}$

(E) 10

GO ON

PSAT Practice Test 1 Section 1 *continued*

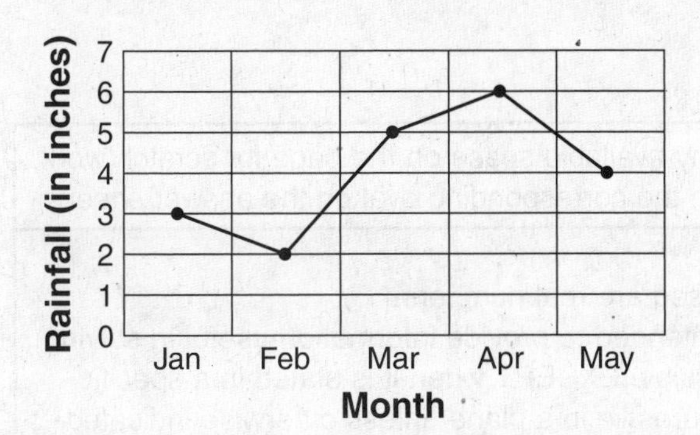

17. The line graph above shows the amount of rainfall in a small town for the first five months of last year. What is the range of the amount of rainfall, in inches, for the five months?

(A) 1

(B) 2

(C) 3

(D) 4

(E) 5

18. Katrina has a collection of 120 animation figurines. If 30 percent of the figurines are of heroines and the rest are of villains, how many villain figurines does she have?

(A) 36

(B) 56

(C) 70

(D) 78

(E) 84

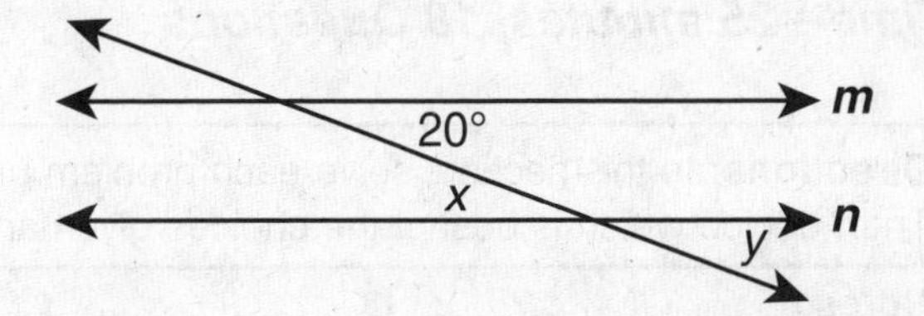

19. Line m is parallel to line n. Angle x and angle y are vertical angles. What is the sum of x and y in degrees?

(A) 20

(B) 40

(C) 160

(D) 180

(E) 320

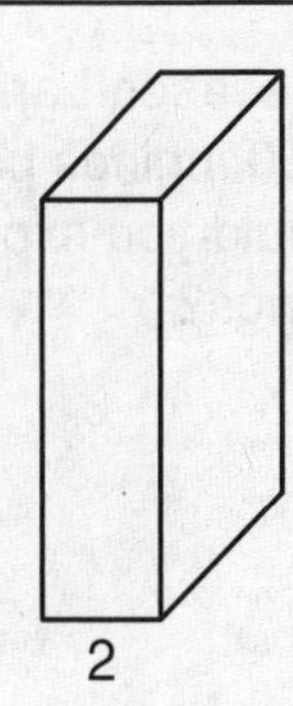

20. The width of the box shown above is 2. If the length of the box is the square of the width and the height is the square of the length, what is the volume of the box in units^3?

(A) 8

(B) 16

(C) 32

(D) 64

(E) 128

STOP **If you finish before time is called, you may check your work on this section only. Do not turn to any other section in the test.**

Name ______________________ Date ____________ Class ____________

PSAT Practice Test 1 Section 2

Time—25 minutes, 18 Questions

Directions: In this section, solve each problem using any available space on the page for scratch work. Then decide which is best of the choices given and fill in the corresponding oval on the answer sheet.

Notes:

1. The use of a calculator is permitted. All numbers used are real numbers.
2. Figures that accompany problems in this test are intended to provide information useful in solving the problems. They are drawn as accurately as possible EXCEPT when it is stated in a specific problem that the figure is not drawn to scale. All figures lie in a plane unless otherwise indicated.

Reference Information

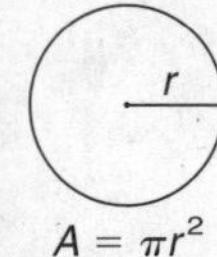

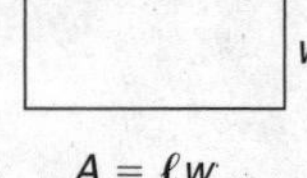
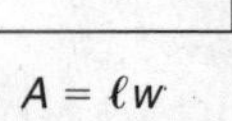
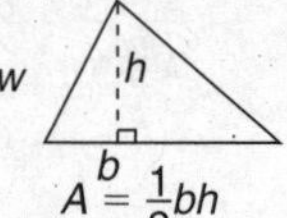
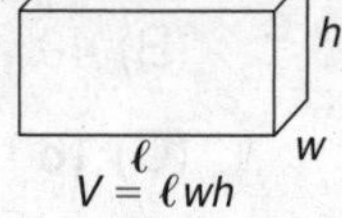
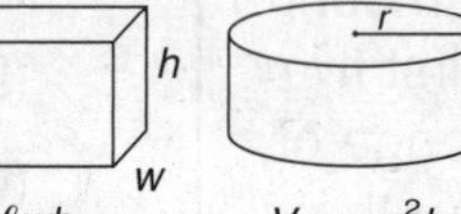
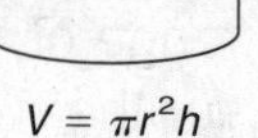
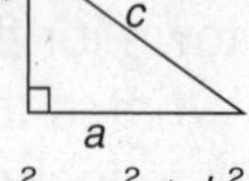
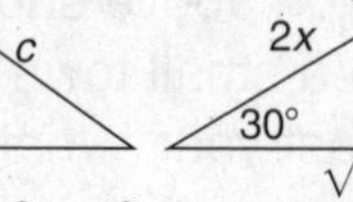
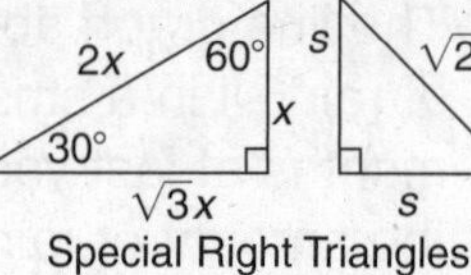

$A = \pi r^2$, $C = 2\pi r$; $A = \ell w$; $A = \frac{1}{2}bh$; $V = \ell wh$; $V = \pi r^2 h$; $c^2 = a^2 + b^2$; Special Right Triangles

The number of degrees of an arc in a circle is 360.
The measure in degrees of a straight angle is 180.
The sum of the measures in degrees of angles of a triangle is 180.

1. This baseball season Larry allowed 100 runs in 360 innings pitched. How many runs would you expect him to allow in 9 innings?

(A) 2

(B) 2.5

(C) 3

(D) 3.5

(E) 4

2. If $x \copyright y = \sqrt{y} + \sqrt{xy}$, then $4 \copyright 9 =$

(A) 6

(B) 8

(C) 9

(D) 10

(E) 18

3. Which of the following is the same as $\frac{1}{\sqrt{4x}} + \frac{\sqrt{x}}{2}$?

(A) $\frac{\sqrt{x} + x\sqrt{x}}{2x}$

(B) $\frac{\sqrt{x} + x}{2x}$

(C) $\frac{1}{2}$

(D) $\frac{\sqrt{x}}{\sqrt{4x} + 2}$

(E) $\frac{1 + \sqrt{x}}{\sqrt{4x} + 2}$

GO ON

Name ______________________ Date __________ Class __________

PSAT Practice Test 1 Section 2 *continued*

4. If $3n - 1$ is an odd integer, what is true about $3n + 1$?

(A) It is an odd integer.

(B) It is an even integer.

(C) It can be an odd or even integer.

(D) It is a prime number.

(E) It is a perfect square.

5. Mike can cut a 40 foot by 20 foot yard with a push mower in 25 minutes. At this rate, how many more minutes would it take him to cut a 40 foot by 30 foot yard?

(A) 12.5

(B) 15

(C) 22.5

(D) 25

(E) 32.5

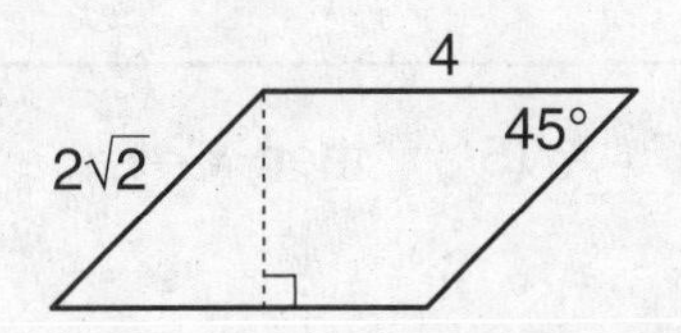

6. What is the area, in square units, of the parallelogram shown above?

(A) 6

(B) 8

(C) 12

(D) 16

(E) Cannot be determined from the information given.

7. If $\sqrt{3x + 4} - 4x = y$ and $x = 3$, what is $|x + y|$?

(A) −5

(B) −4

(C) 2

(D) 4

(E) 5

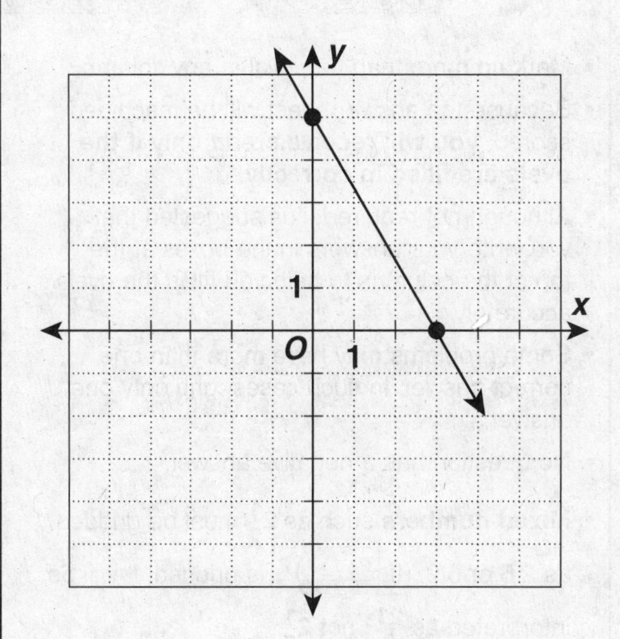

8. If the line shown above is shifted down 5 units and to the right 4 units, what is the x-intercept of the new line?

(A) (−5, 4)

(B) (−4, 0)

(C) (4, 0)

(D) (5, 0)

(E) (5, 4)

If you finish before time is called, you may check your work on this section only. Do not turn to any other section in the test.

Name ______________________ Date __________ Class __________

Directions for Student Response Questions

Each of the remaining 10 questions (9–18) require you to solve the problem and enter your answer by marking the ovals in the special grid, as shown in the examples below.

Answer: $\frac{7}{12}$ or 7/12

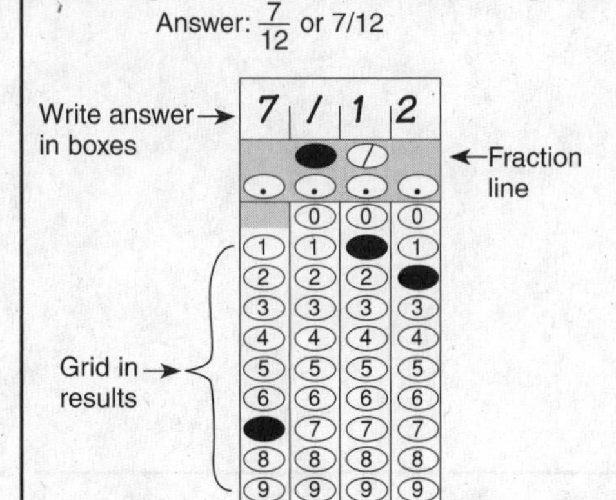

Answer: 2.5

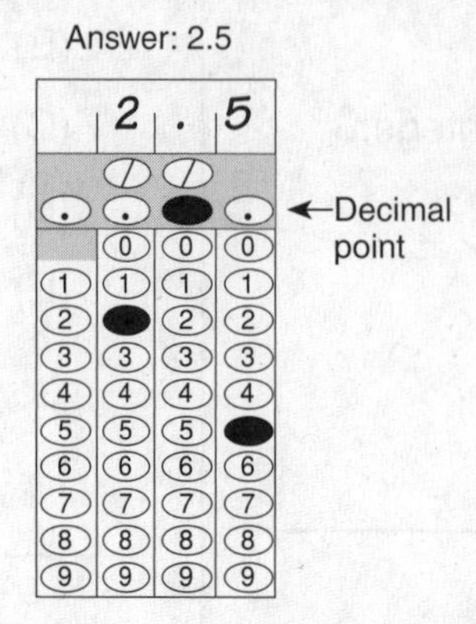

Answer: 201
Either position is correct

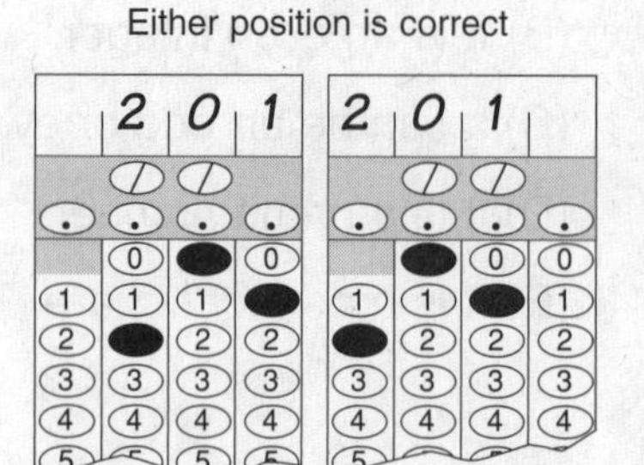

Note: You may start your answers in any column, space permitting. Columns not needed should be left blank

- Mark no more than one oval in any column.
- Because the answer sheet will be machined scored, **you will receive credit only if the ovals are filled in correctly**.
- Although not required, it is suggested that you write your answers in the boxes at the top of the columns to help you fill in the ovals accurately.
- Some problems may have more than one correct answer. In such cases, grid only one answer.
- No question has a negative answer.
- **Mixed numbers** such as $2\frac{1}{2}$ must be gridded as 2.5 or 5/2. (If 2 1 / 2 is gridded, it will be interpreted as $\frac{21}{2}$, not $2\frac{1}{2}$.
- **Decimal Accuracy:** If you obtain a decimal answer, **enter the most accurate value the grid will accommodate**. For example, if you obtain an answer such as 0.6666 …, you should record the result as .666 or .667. **Less accurate values such as .66 or .67 are not acceptable.**

Acceptable ways to grid $\frac{2}{3}$ = .6666 …

2 / 3 | . 6 6 6 | . 6 6 7

9. The first term of a sequence of numbers is −8. Each term after the first is obtained by multiplying the preceding term by −1 and then adding 3. What is the 100th term of the sequence?

10. If $5^{5x-2} = (5^x)^2$, then $x =$

GO ON

PSAT Practice Test 1 Section 2 *continued*

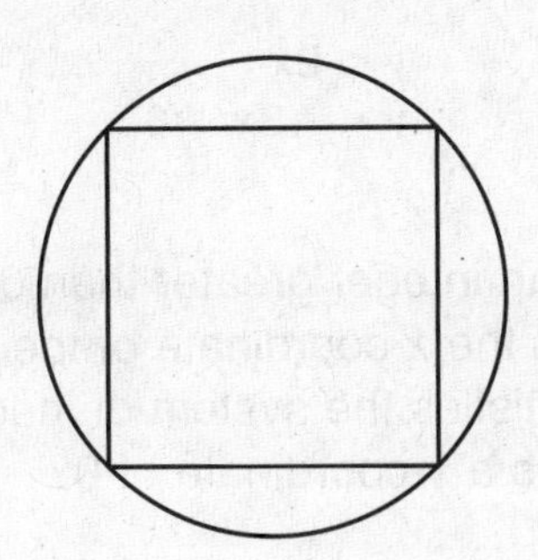

11. The square above is inscribed in the circle. If the length of the diagonal of the square is $\frac{4}{\pi}$, what is the circumference of the circle?

12. If $x^2 + 6x + 9 = 0$, then what is $|2x| - 4$?

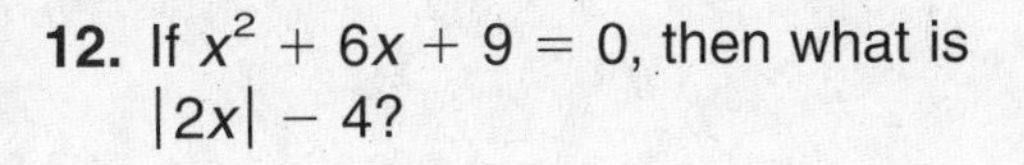

13. Each of the factors of 24 is written on a different piece of paper. If the pieces of paper are placed in a hat and one is drawn at random, what is the probability that the paper has an odd number written on it?

14. The ratio of m to n to p to q is 8 : 5 : 3 : 3. If $p = 21$, what is the value of m?

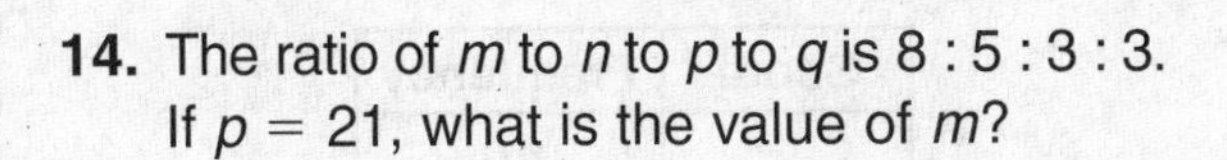

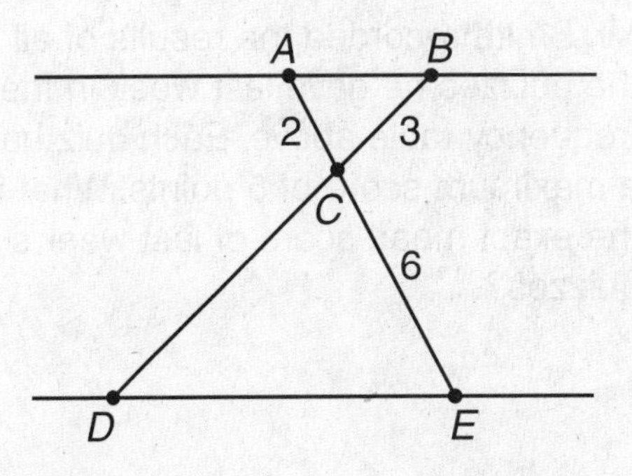

15. If $\overline{AB}$ is parallel to $\overline{DE}$, what is the length of $\overline{CD}$?

16. The area of an equilateral triangle is $16\sqrt{3}$. What is the length of each side of the triangle?

GO ON

Score	Frequency
5	10
4	8
3	12
2	5
1	4
0	1

17. Mr. Smith recorded the results of all the quizzes he gave last week in the frequency table above. Each quiz had a maximum score of 5 points. What is the exact mean score of last week's quizzes?

$y < 2x$
$y < -3x + 6$

18. If x is an integer greater than 0, then what is the x-coordinate of the only point that satisfies the system of inequalities and has a y-coordinate of 0?

STOP **If you finish before time is called, you may check your work on this section only. Do not turn to any other section in the test.**

Name ______________________ Date __________ Class __________

PSAT Practice Test 2 Section 1

Time—25 minutes, 20 Questions

Directions: In this section, solve each problem using any available space on the page for scratch work. Then decide which is best of the choices given and fill in the corresponding oval on the answer sheet.

Notes:

1. The use of a calculator is permitted. All numbers used are real numbers.
2. Figures that accompany problems in this test are intended to provide information useful in solving the problems. They are drawn as accurately as possible EXCEPT when it is stated in a specific problem that the figure is not drawn to scale. All figures lie in a plane unless otherwise indicated.

Reference Information

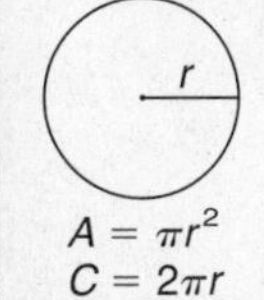

$A = \pi r^2$
$C = 2\pi r$

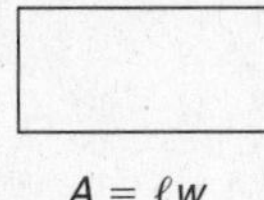

$A = \ell w$

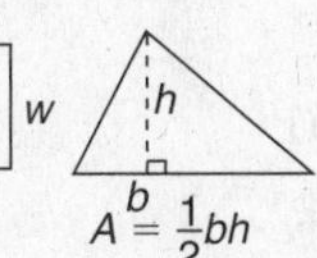

$A = \frac{1}{2}bh$

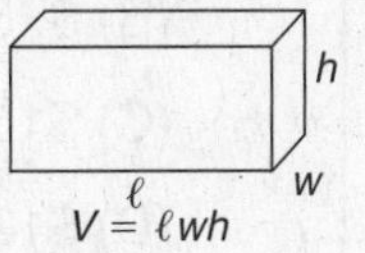

$V = \ell wh$

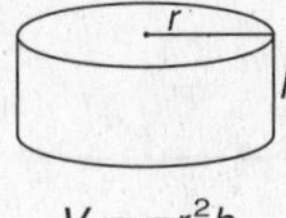

$V = \pi r^2 h$

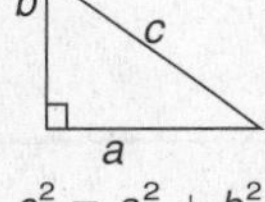

$c^2 = a^2 + b^2$

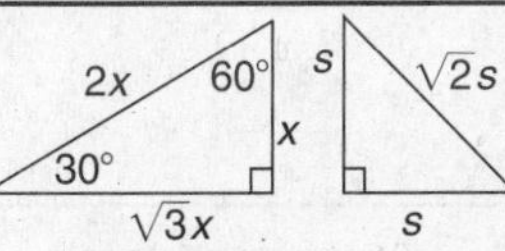

Special Right Triangles

The number of degrees of an arc in a circle is 360.
The measure in degrees of a straight angle is 180.
The sum of the measures in degrees of angles of a triangle is 180.

1. If n is of the form a^2 and a is an integer, which of the following MUST be an odd integer?

(A) $n + 1$

(B) $n + 2$

(C) $2n + 1$

(D) $3n$

(E) $3n + 1$

2. If the measure of one angle of a triangle is equal to the sum of the measures of the other two angles, the triangle is a(n)

(A) equilateral triangle

(B) right triangle

(C) isosceles triangle

(D) obtuse triangle

(E) Cannot be determined from the information given

3. A certain middle school requires that every student participate in at least one of the two science fairs held during the year. If 195 of the 280 students participate in the school fair and 155 participate in the district fair, what percent of the students participate in both fairs?

(A) 20%

(B) 25%

(C) 28%

(D) 34%

(E) 40%

GO ON

PSAT Practice Test 2 Section 1 *continued*

4. If $\frac{4.2}{x} = \frac{4}{5}$, then $\frac{2.1}{x} =$

(A) 0.4

(B) 0.525

(C) 1.4

(D) 2

(E) 5.25

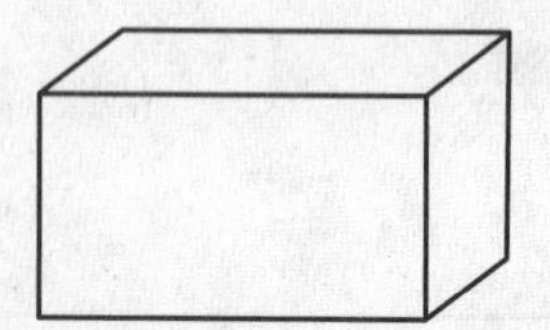

5. If the length, width, and height of the rectangular solid above are integers in the ratio 1 : 2 : 3, which of the following could NOT be the volume of the solid?

(A) 6 cubic units

(B) 48 cubic units

(C) 96 cubic units

(D) 162 cubic units

(E) 384 cubic units

6. Al agreed to clean out his family's basement. He gave 30 percent of his pay to his younger sister Susie for boxing up all the old books. Susie gave 15 percent of her pay to her twin brother Andrew for stacking the boxes after they were full. If Susie gave Andrew $4.50, how much was Al's pay for cleaning the basement?

(A) $45

(B) $75

(C) $100

(D) $125

(E) $150

7. Given the data set {3, 7, 4, 10, 7, 11}, which of the following is NOT equal to the other three?

(A) mean

(B) median

(C) mode

(D) range

(E) All the values are equal.

GO ON

Name ______________________ Date __________ Class __________

8. If $\frac{a}{b+1} = 2$, what does $2b$ equal?

(A) $\frac{a-1}{2}$

(B) $\frac{a-2}{2}$

(C) $a - 1$

(D) $a - 2$

(E) $2a - 1$

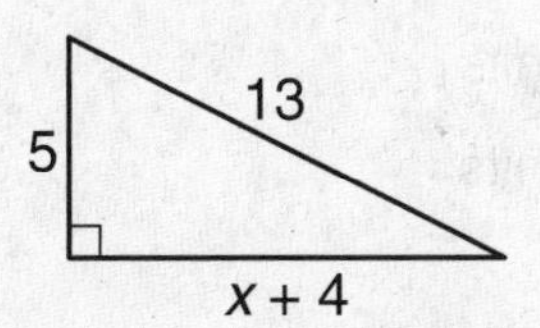

9. What is the value of x in the diagram above?

(A) 4

(B) 5

(C) 8

(D) 11

(E) 12

10. If q is a solution of the equation $3x - 1 = 8$, then what is the value of q^2?

(A) −6

(B) −3

(C) 3

(D) 6

(E) 9

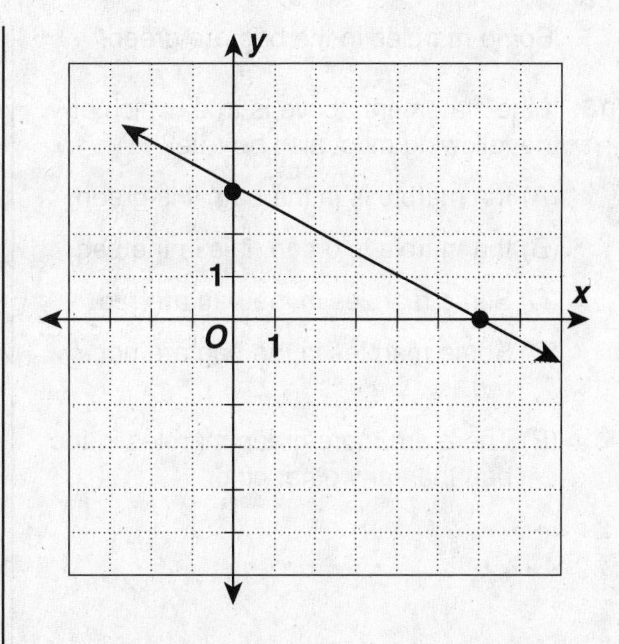

11. The equation of the graph above is

(A) $2x + y = 3$

(B) $x + 2y = 6$

(C) $2x + y = 6$

(D) $-x + 2y = 6$

(E) $-2x + y = 3$

12. If $\frac{1}{3}$ of a number is 4 more than $\frac{1}{4}$ of the number, then the number is

(A) 12

(B) 16

(C) 24

(D) 36

(E) 48

GO ON

"Some marbles in the bag are green."

13. If the statement above is true, which of the following must also be true?

(A) If a marble is in the bag, it is green.

(B) If a marble is green, it is in the bag.

(C) Some marbles in the bag are red.

(D) Some marbles in the bag are not red.

(E) There are more green marbles in the bag than any other color.

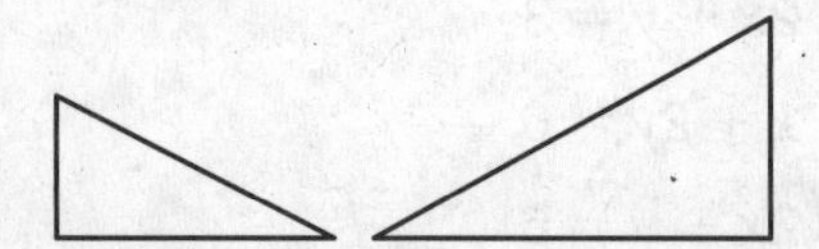

14. The triangles above are similar. If the areas of the two triangles are 16 and 36 respectively, what is the ratio of their perimeters?

(A) 2 : 3

(B) 4 : 9

(C) 9 : 16

(D) 16 : 36

(E) 16 : 81

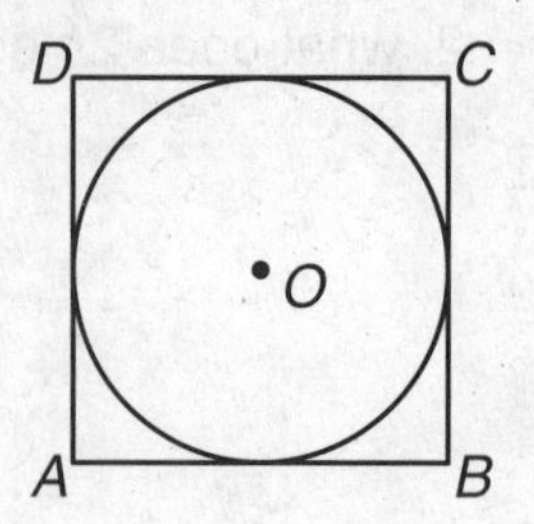

15. Circle O above is inscribed inside square $ABCD$. If the area of the circle is 9π, what is the perimeter of the square?

(A) 6 units

(B) 12 units

(C) 24 units

(D) 36 units

(E) 81 units

16. What is the slope of a line that passes through the origin and the point $(-6, 2)$?

(A) -4

(B) -3

(C) $-\frac{1}{3}$

(D) $\frac{1}{3}$

(E) 3

GO ON

Name ______________________ Date __________ Class __________

PSAT Practice Test 2 Section 1 *continued*

17. If $\frac{a}{b}$ is an integer and $a = 2.4$, which of the following can NOT be the value of b?

(A) $\frac{1}{5}$
(B) 0.8
(C) $\frac{6}{5}$
(D) 2.4
(E) 4.8

18. If $f(x) = \frac{1}{2}x - c$ and $f(8) = 6$, what is the value of c?

(A) -10
(B) -4
(C) -2
(D) 2
(E) 28

19. If $\left(3^{\frac{n}{2}}\right)^{-8} = 3$, what is the value of n?

(A) -1
(B) $-\frac{1}{4}$
(C) 0
(D) 1
(E) 2

9th Grade	4 cans
8th Grade	2 cans
7th Grade	3 cans
6th Grade	2 cans

Legend
1 can = 5 gallons

20. The 6th, 7th, 8th, and 9th grade classes each painted a float for the Thanksgiving parade. Based on the information provided in the chart, which of the following statements is true?

(A) The 9th grade class used two more gallons of paint than the 8th grade.
(B) Together the 6th and 7th grade classes used five gallons of paint.
(C) The 7th grade class used the most paint.
(D) The 7th grade class used five more gallons of paint than the 8th grade class.
(E) The total number of gallons of paint used was 11.

STOP **If you finish before time is called, you may check your work on this section only. Do not turn to any other section in the test.**

Name ____________________ Date __________ Class __________

PSAT Practice Test 2 Section 2

Time—25 minutes, 18 Questions

Directions: In this section, solve each problem using any available space on the page for scratch work. Then decide which is best of the choices given and fill in the corresponding oval on the answer sheet.

Notes:

1. The use of a calculator is permitted. All numbers used are real numbers.
2. Figures that accompany problems in this test are intended to provide information useful in solving the problems. They are drawn as accurately as possible EXCEPT when it is stated in a specific problem that the figure is not drawn to scale. All figures lie in a plane unless otherwise indicated.

Reference Information

$A = \pi r^2$
$C = 2\pi r$

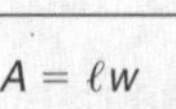

$A = \ell w$

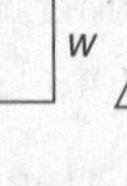

$A = \frac{1}{2}bh$

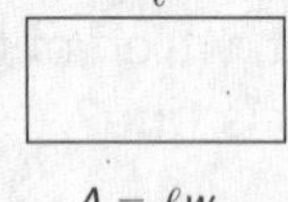

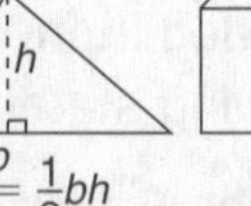

$V = \ell wh$

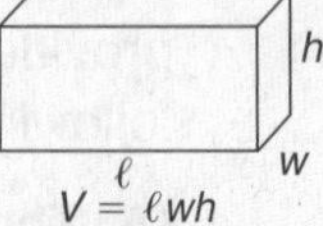

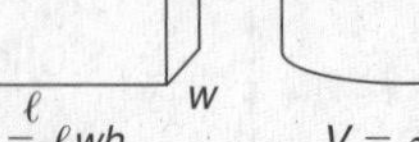

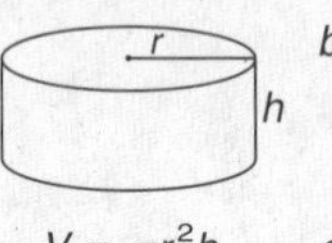

$V = \pi r^2 h$

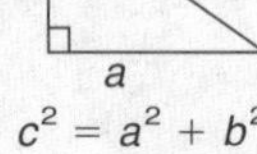

$c^2 = a^2 + b^2$

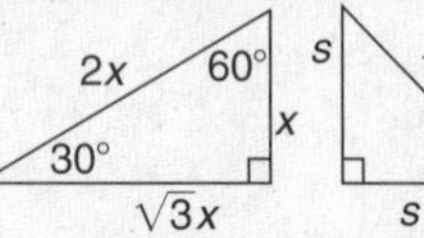

Special Right Triangles

The number of degrees of an arc in a circle is 360.
The measure in degrees of a straight angle is 180.
The sum of the measures in degrees of angles of a triangle is 180.

1. If $2x - y = 7 + y$, then $2y =$

(A) $\frac{7}{2}$

(B) $x - \frac{7}{2}$

(C) $2x - 7$

(D) 14

(E) $x - 14$

2. Which of the following is NOT parallel to $x = -4$?

(A) $x = 2$

(B) $2x = 7$

(C) $x - 1 = 3$

(D) $2x + y = 8 + y$

(E) $x + y = 2y + 4$

3. If point A is reflected through the origin, what is the y-coordinate of the image of A?

(A) −4

(B) −3

(C) 0

(D) 3

(E) 4

GO ON

4. If $-3x \geq 12$, which of the following is true?

(A) $x \leq -4$

(B) $x \geq -4$

(C) $x \leq 4$

(D) $x \leq 15$

(E) $x \geq 15$

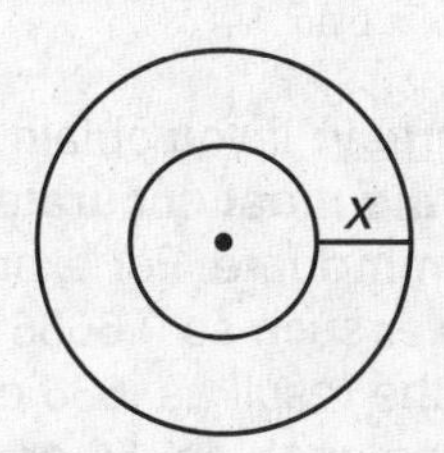

5. If the circumference of the larger circle above is 10π, and the circumference of the smaller circle is 6π, what is the value of x?

(A) 2

(B) 4

(C) 2π

(D) 8

(E) 4π

6. If $a > 0$, $b > 0$, the quotient $\frac{a}{b}$ is 0.25, and $b = a^2$, what is the value of $a + b$?

(A) 6

(B) 8

(C) 12

(D) 16

(E) 20

Joe's Pool Table Sales April	
Table #	**Price**
1	\$2,200
2	\$1,800
3	\$3,000
4	\$2,650
5	\$1,400
6	\$2,775
7	\$3,500

7. Joe's pool table sales for April are recorded in the table above. Joe earns a commission each time he sells a pool table. In any given month, he earns 5% commission on each of the first three pool tables he sells. He earns 6% commission on the fourth table he sells, 7% on the fifth table, and so on up to a maximum of 20%. What is Joe's total commission in April?

(A) \$866.25

(B) \$937.50

(C) \$1021.00

(D) \$1144.00

(E) \$1559.25

8. If $3|x| = \frac{1}{2}$, which of the following could be a value of x?

(A) -6

(B) $-\frac{3}{2}$

(C) $-\frac{1}{6}$

(D) $\frac{3}{2}$

(E) 6

GO ON

PSAT Practice Test 2 Section 2 *continued*

Directions for Student Response Questions

Each of the remaining 10 questions (9–18) require you to solve the problem and enter your answer by marking the ovals in the special grid, as shown in the examples below.

Answer: $\frac{7}{12}$ or 7/12

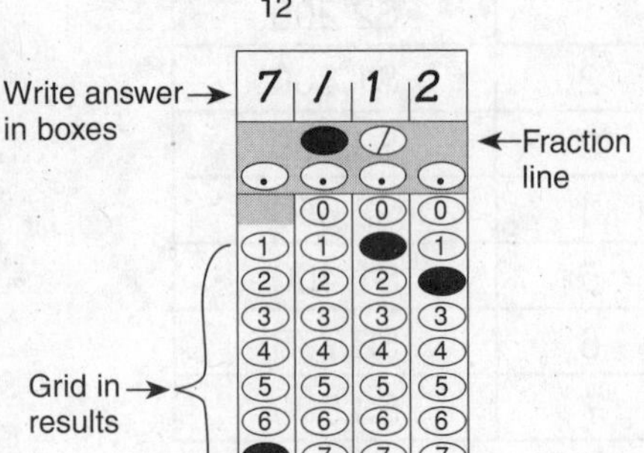

Answer: 2.5

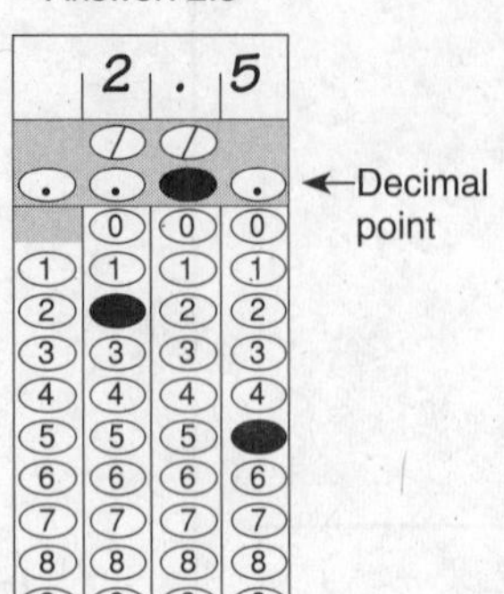

Answer: 201
Either position is correct

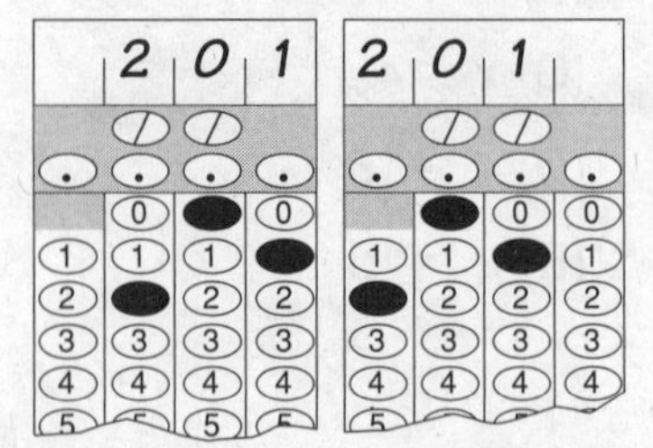

Note: You may start your answers in any column, space permitting. Columns not needed should be left blank

- Mark no more than one oval in any column.
- Because the answer sheet will be machined scored, **you will receive credit only if the ovals are filled in correctly**.
- Although not required, it is suggested that you write your answers in the boxes at the top of the columns to help you fill in the ovals accurately.
- Some problems may have more than one correct answer. In such cases, grid only one answer.
- No question has a negative answer.
- **Mixed numbers** such as $2\frac{1}{2}$ must be gridded as 2.5 or 5/2. (If 2 1 / 2 is gridded, it will be interpreted as $\frac{21}{2}$, not $2\frac{1}{2}$.
- **Decimal Accuracy:** If you obtain a decimal answer, **enter the most accurate value the grid will accommodate**. For example, if you obtain an answer such as 0.6666 …, you should record the result as .666 or .667. **Less accurate values such as .66 or .67 are not acceptable.**

Acceptable ways to grid $\frac{2}{3}$ = .6666 …

2 / 3   . 6 6 6   . 6 6 7

9. If y is indirectly proportional to x and if $y = 12$ when $x = 3$, what is the value of y when $x = 4$?

10. The mean of the set { 2, 5, x, 9, 13 } is 7. What is the median?

GO ON

Name ______________________ Date __________ Class __________

PSAT Practice Test 2 Section 2 *continued*

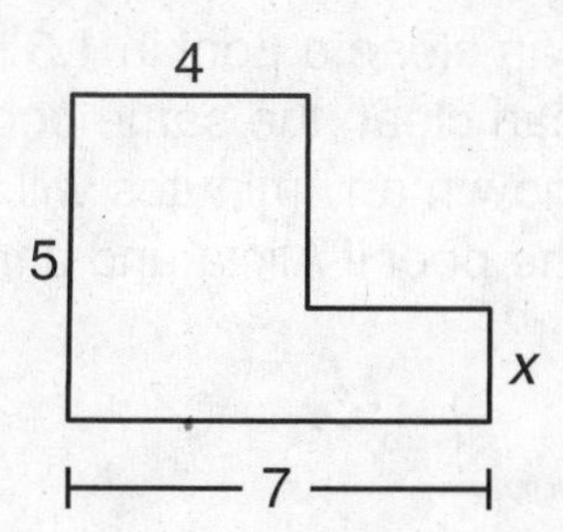

11. If the area of the polygon above is 26 square units, what is the value of x?

12. If 20% of one number is 8, and 20% of another number is 9, what is the sum of the two numbers?

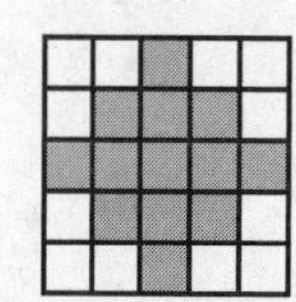

13. Each of the smaller squares inside the larger square shown above is the same size. If the diagram represents a dart board, what is the difference between the probability of hitting a shaded square and the probability of hitting an unshaded square, assuming the board is hit?

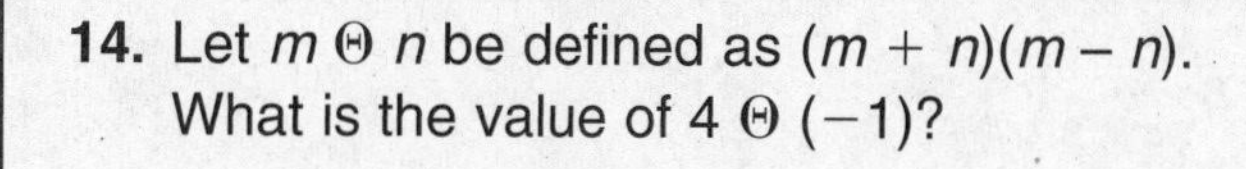

14. Let $m \ominus n$ be defined as $(m + n)(m - n)$. What is the value of $4 \ominus (-1)$?

15. A certain air compressor can fill 75 balloons in 15 minutes. At this rate, how many more minutes would it take to fill 200 balloons?

GO ON

PSAT Practice Test 2 Section 2 *continued*

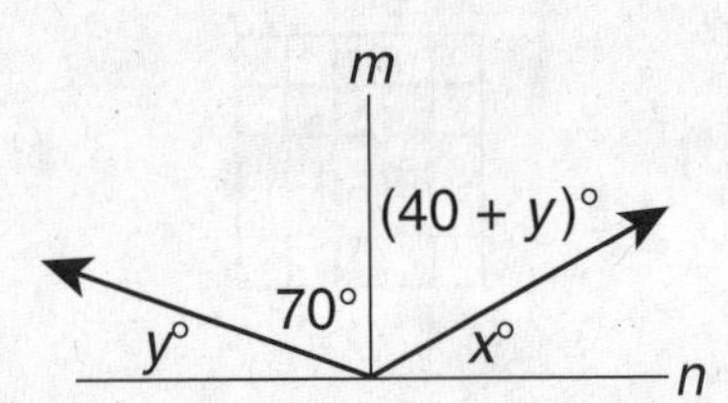

16. If line n is perpendicular to line m in the diagram above, what is the value of x in degrees?

17. If a and b are roots of the equation $x^2 - 13x - 30$, and $a > b$, what is the value of $a - b$?

18. Alicia can clean a pool in 1.5 hours. Jared can clean the same pool in 1 hour. About how many minutes will it take to clean the pool if Alicia and Jared work together?

STOP **If you finish before time is called, you may check your work on this section only. Do not turn to any other section in the test.**

Name ______________________ Date __________ Class __________

SAT Practice Test

Completely darken bubbles with a No. 2 pencil. If you make a mistake, be sure to erase mark completely. Erase all stray marks.

1. YOUR NAME: ______________________
(Print) Last First M.I.

SIGNATURE: ______________ **DATE:** __________

HOME ADDRESS: ______________________
(Print) Number and Street

______________________ **E-MAIL:** __________
City State Zip

PHONE NO.: __________ **SCHOOL:** __________ **CLASS OF:** __________
(Print)

IMPORTANT: Please fill in these boxes exactly as shown on the back cover of your text book.

2. TEST FORM

3. TEST CODE

0	0	0	0
1	1	1	1
2	2	2	2
3	3	3	3
4	4	4	4
5	5	5	5
6	6	6	6
7	7	7	7
8	8	8	8
9	9	9	9

4. PHONE NUMBER

0	0	0	0	0	0	0
1	1	1	1	1	1	1
2	2	2	2	2	2	2
3	3	3	3	3	3	3
4	4	4	4	4	4	4
5	5	5	5	5	5	5
6	6	6	6	6	6	6
7	7	7	7	7	7	7
8	8	8	8	8	8	8
9	9	9	9	9	9	9

5. YOUR NAME

First 4 letters of last name				FIRST INT	LAST INT
A	A	A	A	A	A
B	B	B	B	B	B
C	C	C	C	C	C
D	D	D	D	D	D
E	E	E	E	E	E
F	F	F	F	F	F
G	G	G	G	G	G
H	H	H	H	H	H
I	I	I	I	I	I
J	J	J	J	J	J
K	K	K	K	K	K
L	L	L	L	L	L
M	M	M	M	M	M
N	N	N	N	N	N
O	O	O	O	O	O
P	P	P	P	P	P
Q	Q	Q	Q	Q	Q
R	R	R	R	R	R
S	S	S	S	S	S
T	T	T	T	T	T
U	U	U	U	U	U
V	V	V	V	V	V
W	W	W	W	W	W
X	X	X	X	X	X
Y	Y	Y	Y	Y	Y
Z	Z	Z	Z	Z	Z

6. DATE OF BIRTH

MONTH	DAY		YEAR	
JAN	0	0	0	0
FEB	1	1	1	1
MAR	2	2	2	2
APR	3	3	3	3
MAY		4	4	4
JUN		5	5	5
JUL		6	6	6
AUG		7	7	7
SEP		8	8	8
OCT		9	9	9
NOV				
DEC				

7. SEX

- MALE
- FEMALE

8. OTHER

1	A	B	C	D	E
2	A	B	C	D	E
3	A	B	C	D	E

1 MATHEMATICS

1	A B C D E	6	A B C D E	11	A B C D E	16	A B C D E
2	A B C D E	7	A B C D E	12	A B C D E	17	A B C D E
3	A B C D E	8	A B C D E	13	A B C D E	18	A B C D E
4	A B C D E	9	A B C D E	14	A B C D E	19	A B C D E
5	A B C D E	10	A B C D E	15	A B C D E	20	A B C D E

Name ____________________ Date __________ Class __________

SAT Practice Test

Use a No. 2 pencil only. Be sure each mark is dark and completely fills the intended oval. Completely erase any errors o stray marks.

2 MATHEMATICS

1 (A)(B)(C)(D)(E)
2 (A)(B)(C)(D)(E)
3 (A)(B)(C)(D)(E)
4 (A)(B)(C)(D)(E)
5 (A)(B)(C)(D)(E)
6 (A)(B)(C)(D)(E)
7 (A)(B)(C)(D)(E)
8 (A)(B)(C)(D)(E)

ONLY ANSWERS ENTERED IN THE OVALS IN EACH GRID AREA WILL BE SCORED. YOU WILL NOT RECEIVE CREDIT FOR ANYTHING WRITTEN IN THE BOXES ABOVE THE OVALS.

9

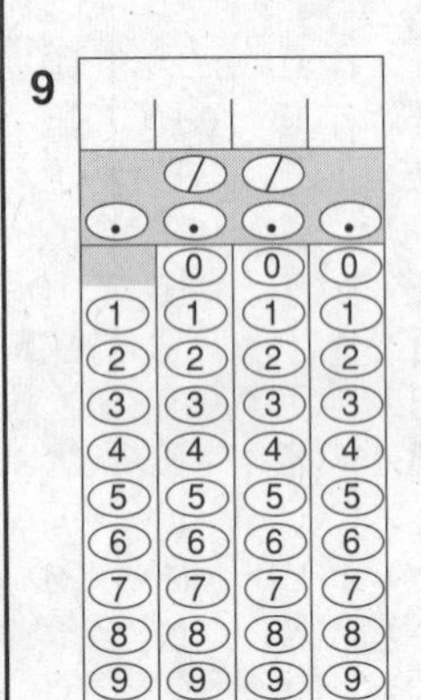

10

11

12

13

14

15

16

17

18

3 MATHEMATICS

1 (A)(B)(C)(D)(E)
2 (A)(B)(C)(D)(E)
3 (A)(B)(C)(D)(E)
4 (A)(B)(C)(D)(E)
5 (A)(B)(C)(D)(E)
6 (A)(B)(C)(D)(E)
7 (A)(B)(C)(D)(E)
8 (A)(B)(C)(D)(E)
9 (A)(B)(C)(D)(E)
10 (A)(B)(C)(D)(E)
11 (A)(B)(C)(D)(E)
12 (A)(B)(C)(D)(E)
13 (A)(B)(C)(D)(E)
14 (A)(B)(C)(D)(E)
15 (A)(B)(C)(D)(E)
16 (A)(B)(C)(D)(E)

Name ____________________ Date __________ Class __________

SAT Practice Test 1 Section 1

Time—25 minutes, 20 Questions

Directions: In this section, solve each problem using any available space on the page for scratch work. Then decide which is best of the choices given and fill in the corresponding oval on the answer sheet.

Notes:

1. The use of a calculator is permitted. All numbers used are real numbers.
2. Figures that accompany problems in this test are intended to provide information useful in solving the problems. They are drawn as accurately as possible EXCEPT when it is stated in a specific problem that the figure is not drawn to scale. All figures lie in a plane unless otherwise indicated.

Reference Information

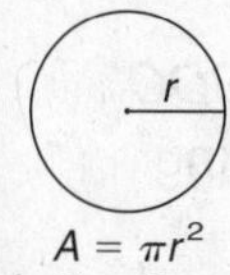

$A = \pi r^2$
$C = 2\pi r$

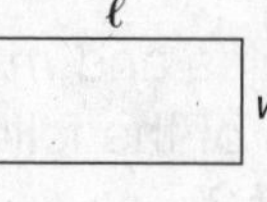

$A = \ell w$

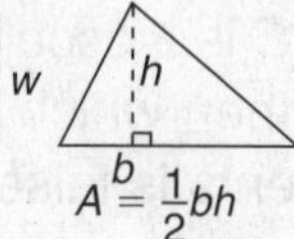

$A = \frac{1}{2}bh$

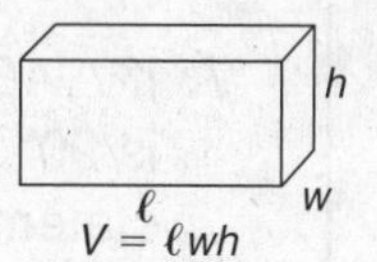

$V = \ell wh$

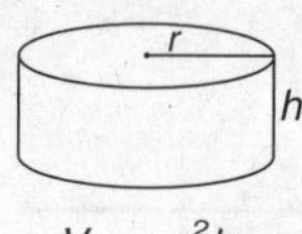

$V = \pi r^2 h$

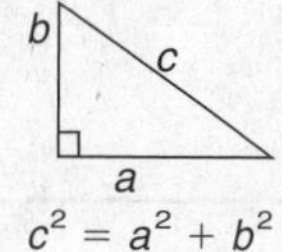

$c^2 = a^2 + b^2$

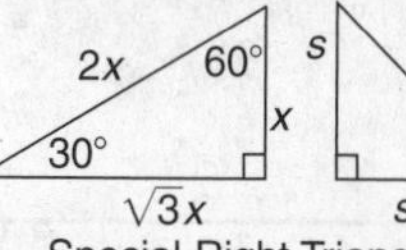

Special Right Triangles

The number of degrees of an arc in a circle is 360.
The measure in degrees of a straight angle is 180.
The sum of the measures in degrees of angles of a triangle is 180.

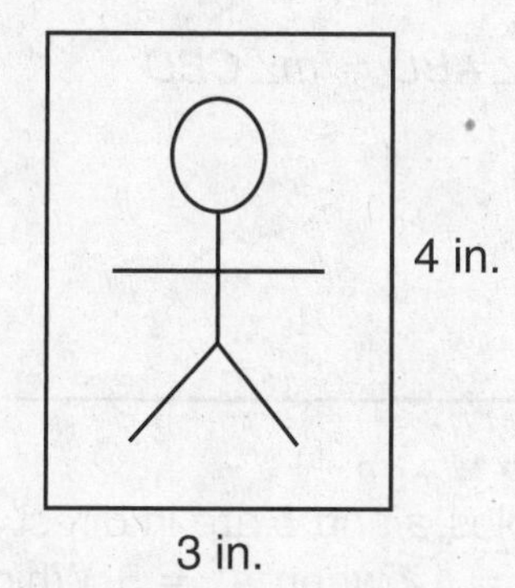

1. If the picture shown above is enlarged proportionally so that the height is now 6 inches, how large of a border would you need so that it would go all the way around the enlarged picture?

(A) 13 inches

(B) 15 inches

(C) 17 inches

(D) 19 inches

(E) 21 inches

2. If a circle with radius 5 has its center at the point $(-1, 3)$, which of the following points is on the circle?

(A) $(-6, -2)$

(B) $(-1, -2)$

(C) $(-4, 8)$

(D) $(6, 3)$

(E) $(4, 8)$

3. The graph of which of the following equations would be perpendicular to the graph of $y = -3x + 8$?

(A) $12x + 36y = 1$

(B) $3x - y = 8$

(C) $3x - 8y = 1$

(D) $3x + y = 8$

(E) $4x - 12y = 7$

GO ON

SAT Practice Test 1 Section 1 *continued*

4. If $\frac{7.2}{x} = \frac{4}{5}$, then $\frac{x}{4} =$

(A) 0.25
(B) 1.25
(C) 2.25
(D) 3
(E) 9

5. If Jane goes to the market, she cannot take Pete to baseball practice. If Pete goes to practice, he can play in the game tomorrow. If Pete does not play in tomorrow's game, then which of the following MUST be true?

(A) Jane went to the market.
(B) Jane did not go to the market.
(C) Pete went to practice.
(D) Pete did not go to practice.
(E) None of the above.

6. $3x + 2y = 10$
$2x + y = 7$

Use the system given above to find the sum of x and y.

(A) 3
(B) 4
(C) 5
(D) 6
(E) 7

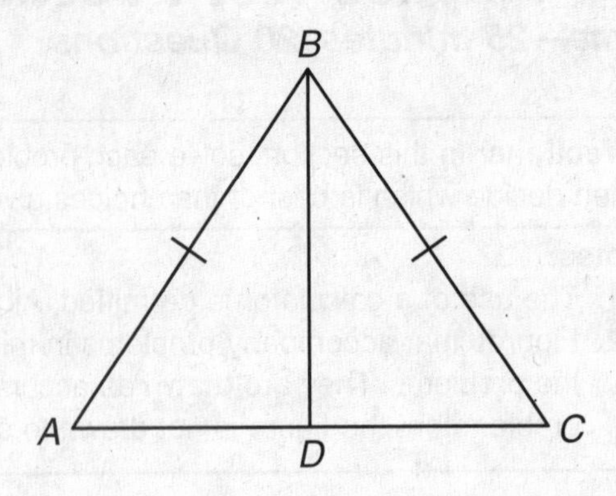

7. If $\triangle ABC$ is isosceles and $m\angle BDC$ is 90°, then which of the following statements is false?

(A) $BD = AC$
(B) $AD = DC$
(C) $AB = BC$
(D) $m\angle A = m\angle C$
(E) $m\angle ABD = m\angle CBD$

8. Variables a and b are in direct variation and $b = 12$ when $a^2 = 5$. Which of the following could be the value of a when $b = 6$?

(A) $-\frac{5}{2}$
(B) -2
(C) $-\frac{\sqrt{5}}{2}$
(D) 2
(E) $\frac{5}{2}$

GO ON

9. If a, b, c, and d are constants, which of the following does NOT represent a function in terms of x?

(A) $x = a$

(B) $y = ax + b$

(C) $x = ay^2 + b$

(D) $y = ax^2 + bx + c$

(E) $y = (x - a)^2 + b$

---

10. Which of the following statements is ALWAYS true for a set of data?

(A) The mode is greater than the range.

(B) The median is greater than the mean.

(C) The mean cannot equal the median.

(D) The mode always has the greatest value.

(E) None of the above.

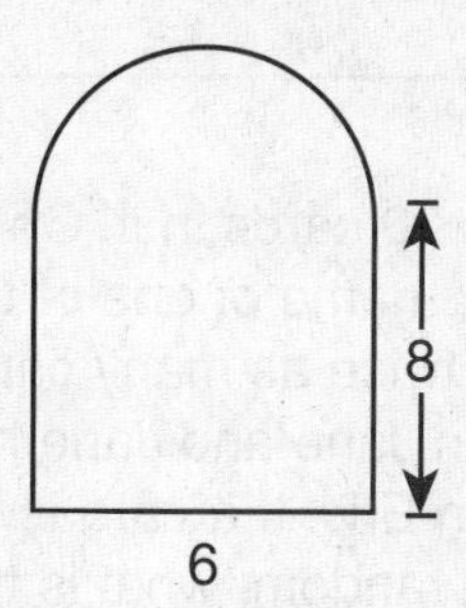

11. The shape above was made by connecting a semicircle to a rectangle. What is the perimeter of the shape?

(A) $3\pi + 22$

(B) $6\pi + 14$

(C) $6\pi + 22$

(D) $12\pi + 22$

(E) 25π

12. If m and n are roots of the equation $x^2 - 9x + 20 = 0$, what is the value of $m^2 + n^2$?

(A) 9

(B) 12

(C) 41

(D) 81

(E) 104

---

13. Shawna is at the state fair. She currently has enough money to ride the Berserker 20 times. If the cost to ride the Berserker was 25 cents less, Shawna could ride it 10 more times. How much money does Shawna have?

(A) $10.00

(B) $15.00

(C) $17.00

(D) $20.00

(E) $23.00

GO ON

SAT Practice Test 1 Section 1 *continued*

14. If $x + y = 1.2$, then $x^2 + 2xy + y^2 =$

(A) 1.44

(B) 2.4

(C) 3.84

(D) 3.6

(E) Cannot be determined from the information given

$$\frac{1}{2}, \ ?, \ \frac{25}{18}, \ -\frac{125}{54}, \ldots$$

15. What number is missing from the sequence shown above?

(A) $-\frac{12}{10}$

(B) $-\frac{5}{6}$

(C) $\frac{5}{12}$

(D) $\frac{5}{6}$

(E) $\frac{12}{10}$

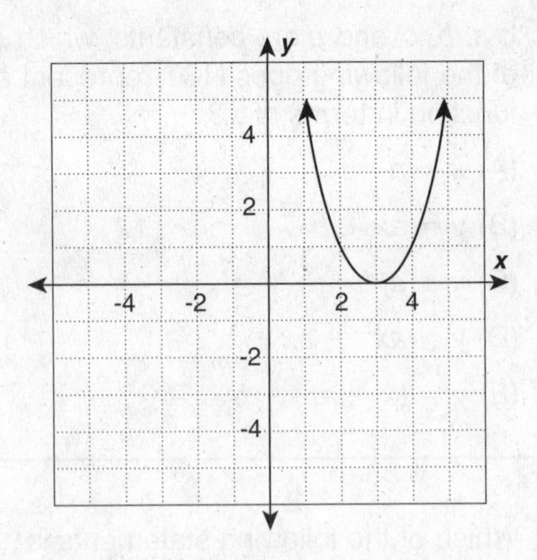

16. The graph of $f(x)$ is shown above. If $f(x + 4)$ were graphed instead, where would the vertex of the parabola be?

(A) $(0, -1)$

(B) $(-1, 0)$

(C) $(0, 7)$

(D) $(7, 0)$

(E) $(-1, 7)$

17. A hat has 20 cards in it. On each card is written the name of one of three people. Mike has twice as many cards with his name than Jane and Jane has 4 more cards than Eric. If a card is pulled out of the hat at random, what is the probability that it has Eric's name on it?

(A) $\frac{1}{20}$

(B) $\frac{1}{10}$

(C) $\frac{1}{6}$

(D) $\frac{1}{3}$

(E) $\frac{1}{2}$

GO ON

SAT Practice Test 1 Section 1 *continued*

18. A soup can is made up of a side, a top, and a bottom. If the diameter and height of the can are equal and the volume is 128π units^3, what is the total surface area of the can in square units?

(A) 84π

(B) 96π

(C) 128π

(D) 140π

(E) 152π

19. The length of a rectangle is twice the width. If the area of the rectangle is 32 square units, what is the perimeter of the rectangle?

(A) 4

(B) 8

(C) 16

(D) 24

(E) 32

20. If $a \Phi b = \frac{a+b}{a-b}$, then $\frac{1}{2} \Phi 1 = ?$

(A) -3

(B) -1

(C) 1

(D) $\frac{3}{2}$

(E) 3

STOP **If you finish before time is called, you may check your work on this section only. Do not turn to any other section in the test.**

Name ________________ Date ________ Class ________

SAT Practice Test 1 Section 2

Time—25 minutes, 18 Questions

Directions: In this section, solve each problem using any available space on the page for scratch work. Then decide which is best of the choices given and fill in the corresponding oval on the answer sheet.

Notes:

1. The use of a calculator is permitted. All numbers used are real numbers.
2. Figures that accompany problems in this test are intended to provide information useful in solving the problems. They are drawn as accurately as possible EXCEPT when it is stated in a specific problem that the figure is not drawn to scale. All figures lie in a plane unless otherwise indicated.

Reference Information

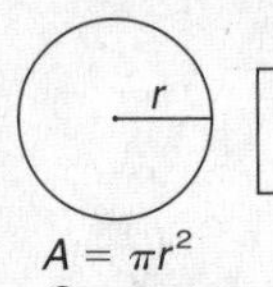
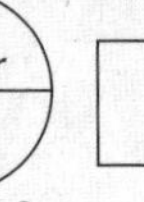

$A = \pi r^2$
$C = 2\pi r$

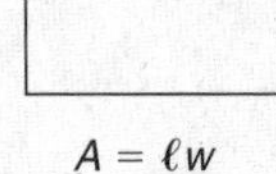

$A = \ell w$

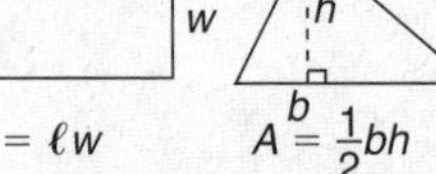

$A = \frac{1}{2}bh$

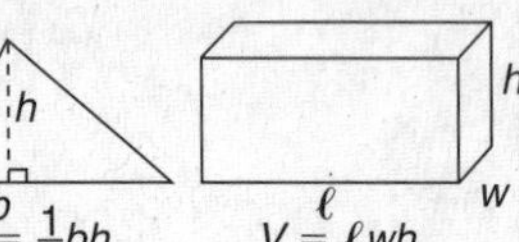

$V = \ell wh$

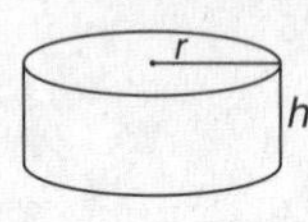
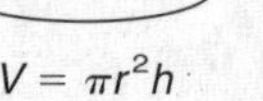

$V = \pi r^2 h$

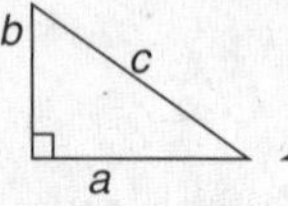

$c^2 = a^2 + b^2$

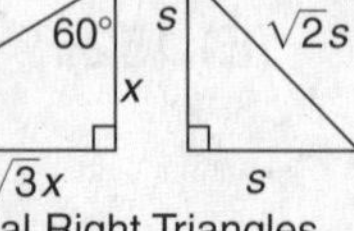
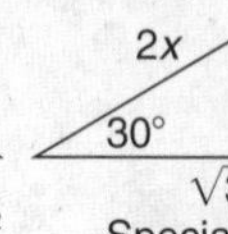

Special Right Triangles

The number of degrees of an arc in a circle is 360.
The measure in degrees of a straight angle is 180.
The sum of the measures in degrees of angles of a triangle is 180.

1. Mary bought several boxes of mechanical pencils. Each box contained 12 pencils. Mary sold some of the pencils to her friends and kept the others. If she sold 25 percent more than she kept, and she sold 80 pencils, how many boxes did she buy?

(A) 8

(B) 9

(C) 10

(D) 11

(E) 12

2. What is the value of a so that the line that passes through the points $(a, 7)$ and $(5, 4)$ is perpendicular to the line $y = 2x + 7$?

(A) –2

(B) –1

(C) 0

(D) 1

(E) 2

$$x - y \geq 0$$
$$2x - 7y \leq 0$$

3. Which of the following is a member of the solution set to the system of inequalities shown above?

(A) the point $(-1, -2)$

(B) the line $x = 2$

(C) the line $y = 0$

(D) the origin

(E) the point $(-7, -2)$

GO ON

SAT Practice Test 1 Section 2 *continued*

4. Mary is given five tests that all have a maximum score of 100 points. The median score of her tests is 72. What is the best mean score, rounded to the nearest whole number, that she could have earned?

(A) 72

(B) 83

(C) 86

(D) 88

(E) 90

5. Bill and Ross are twins. Their brother Mark was born on the same day as the twins, but is three years older. If Mark was born in 1990, in what year did the three boys' ages total 21?

(A) 1995

(B) 1996

(C) 1997

(D) 1998

(E) 1999

6. The line $2x + 3y = 5$ is parallel to the line that passes through which two points?

(A) $(-2, 0)$ and $(0, -3)$

(B) $(-1, 1)$ and $(2, 6)$

(C) $(1, 1)$ and $(-2, 3)$

(D) $(1, 1)$ and $(3, -2)$

(E) $(6, 4)$ and $(10, -2)$

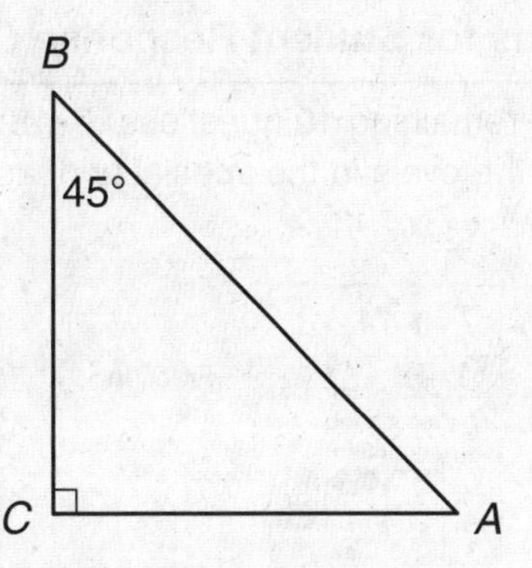

7. The ratio $AB : BC$ equals

(A) $\frac{1}{\sqrt{3}}$

(B) $\frac{1}{\sqrt{2}}$

(C) $\frac{\sqrt{2}}{\sqrt{3}}$

(D) $\sqrt{2}$

(E) $\sqrt{3}$

8. Solve the equation for x: $y = mx + b$.

(A) $x = y - m - b$

(B) $x = \frac{y}{m} - b$

(C) $x = my + b$

(D) $x = y + m + b$

(E) $x = \frac{y - b}{m}$

GO ON

Name ______________________ Date __________ Class __________

SAT Practice Test 1 Section 2 *continued*

Directions for Student Response Questions

Each of the remaining 10 questions (9–18) require you to solve the problem and enter your answer by marking the ovals in the special grid, as shown in the examples below.

Answer: $\frac{7}{12}$ or 7/12

Write answer in boxes → 7 / 1 2 ← Fraction line

Grid in results

Answer: 2.5

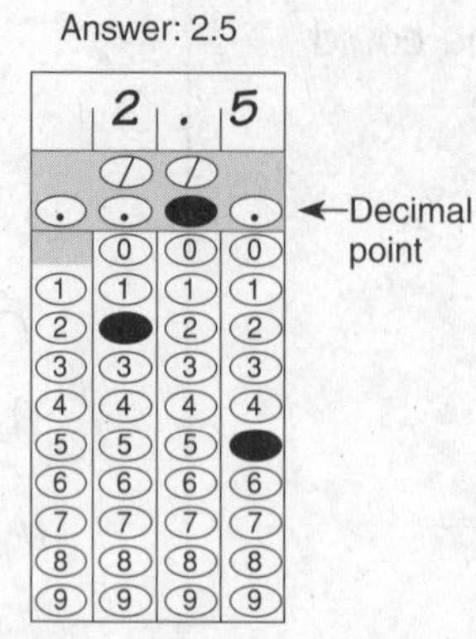

← Decimal point

Answer: 201
Either position is correct

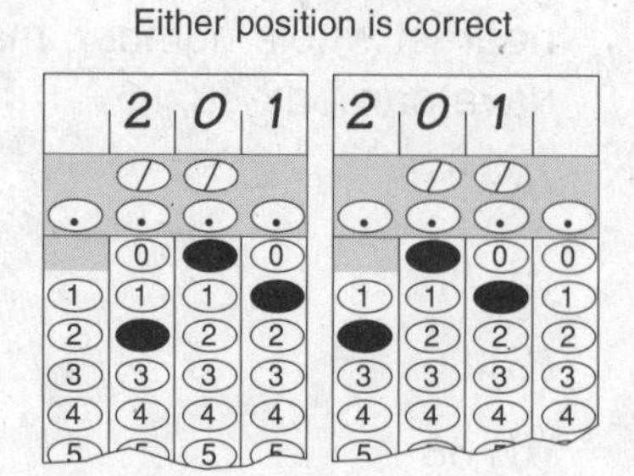

Note: You may start your answers in any column, space permitting. Columns not needed should be left blank

- Mark no more than one oval in any column.
- Because the answer sheet will be machined scored, **you will receive credit only if the ovals are filled in correctly**.
- Although not required, it is suggested that you write your answers in the boxes at the top of the columns to help you fill in the ovals accurately.
- Some problems may have more than one correct answer. In such cases, grid only one answer.
- No question has a negative answer.
- **Mixed numbers** such as $2\frac{1}{2}$ must be gridded as 2.5 or 5/2. (If 2 1 / 2 is gridded, it will be interpreted as $\frac{21}{2}$, not $2\frac{1}{2}$.
- **Decimal Accuracy:** If you obtain a decimal answer, **enter the most accurate value the grid will accommodate.** For example, if you obtain an answer such as 0.6666 …, you should record the result as .666 or .667. **Less accurate values such as .66 or .67 are not acceptable.**

Acceptable ways to grid $\frac{2}{3}$ = .6666 …

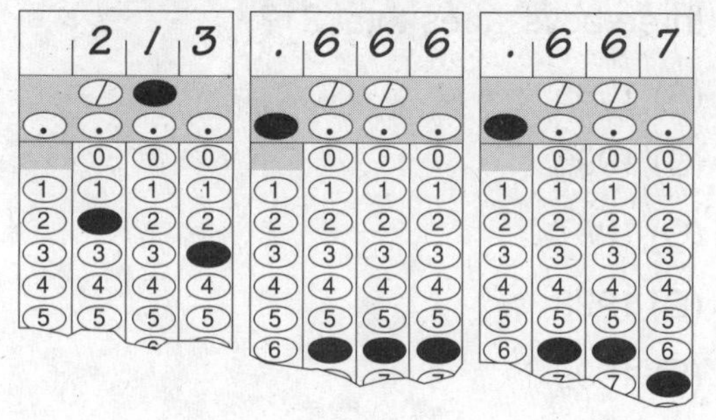

9. Sean has a collection of 48 lighthouse replicas. Seventy-five percent of his collection are replicas of lighthouses found on the east coast of the United States and 75 percent of those are from the New England states. How many of the replicas in Sean's collection are from the east coast but NOT from the New England states?

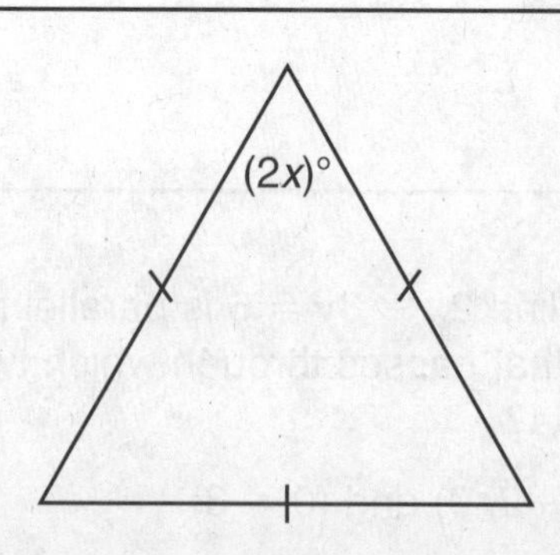

10. What is the value of x in the diagram above?

GO ON

SAT Practice Test 1 Section 2 *continued*

11. According to the map, the gas station is 5 km west and 2 km south of your house. The post office is 3 km east and 4 km north of your house. How many kilometers apart are the gas station and the post office?

12. If $5^{\frac{n}{2}} + 5^{\frac{n}{2}} + 5^{\frac{n}{2}} + 5^{\frac{n}{2}} + 5^{\frac{n}{2}} = 5^{\frac{n}{2}+a}$, what is the value of a?

13. A gas tank is filled $\frac{3}{4}$ of the way to the top. Joe needs $\frac{1}{2}$ of the gas and Mary needs $\frac{1}{3}$ of the gas that is in the tank. What fraction of the tank is filled with gas after Joe and Mary take what they need?

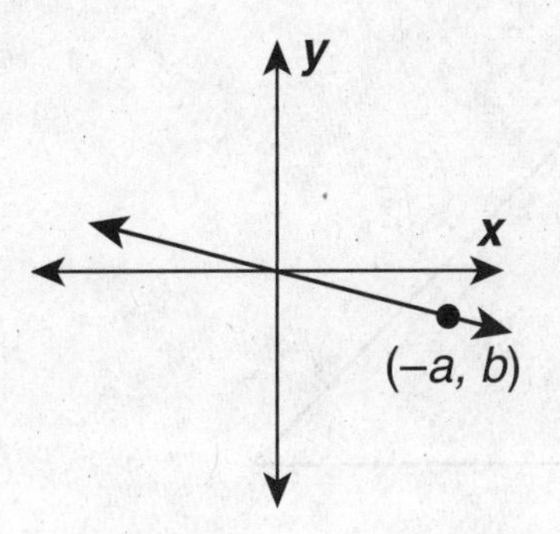

14. If the slope of the line shown above is $-\frac{2}{7}$ and $a = 14$, what is the absolute value of b?

15. If $2\sqrt{xa^4} + a\sqrt{2a^2} = 7a^2\sqrt{2}$, then $x =$

GO ON

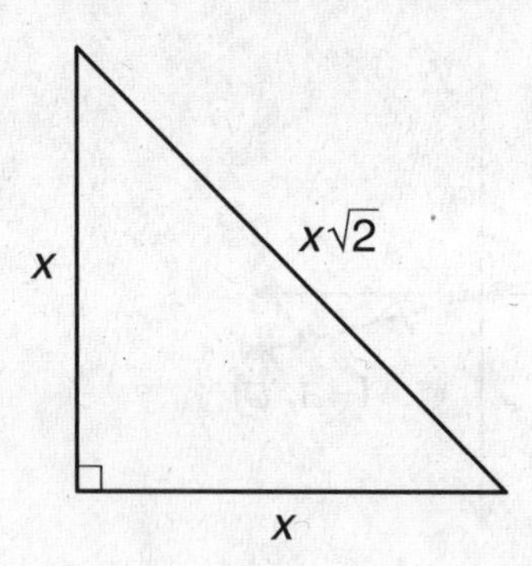

16. The perimeter of the right triangle shown above is $14 + 7\sqrt{2}$ units. What is the area of the triangle in square units?

17. A set of data has a minimum value of -2 and a range of 16. A second set of data has a minimum value of 5 and a range of 16. What is the positive difference between the maximum values of the two sets of data?

18. If $4^{x+8} = 64^{x}$, then $x =$

STOP **If you finish before time is called, you may check your work on this section only. Do not turn to any other section in the test.**

Name ______________________ Date __________ Class __________

SAT Practice Test 1 Section 3

Time—20 minutes, 16 Questions

Directions: In this section, solve each problem using any available space on the page for scratch work. Then decide which is best of the choices given and fill in the corresponding oval on the answer sheet.

Notes:

1. The use of a calculator is permitted. All numbers used are real numbers.
2. Figures that accompany problems in this test are intended to provide information useful in solving the problems. They are drawn as accurately as possible EXCEPT when it is stated in a specific problem that the figure is not drawn to scale. All figures lie in a plane unless otherwise indicated.

Reference Information

$A = \pi r^2$, $C = 2\pi r$ $A = \ell w$ $A = \frac{1}{2}bh$ $V = \ell wh$ $V = \pi r^2 h$ $c^2 = a^2 + b^2$

Special Right Triangles: $2x$, $60°$, x, $30°$, $\sqrt{3}x$; s, $\sqrt{2}s$, s

The number of degrees of an arc in a circle is 360.
The measure in degrees of a straight angle is 180.
The sum of the measures in degrees of angles of a triangle is 180.

Letter	Amount
A	16
B	12
E	8
G	
O	

1. There are 100 cards in a hat and each card has one of five letters on it. Bekah has not finished making the table above that shows how many of each letter is in the hat. What is the maximum amount of vowels in the hat? Assume that each letter appears at least once.

(A) 16
(B) 24
(C) 52
(D) 64
(E) 87

2. Marie won $1,000,000. Each day she spends half of the money that she had the day before. If at the end of day 1 she has the million dollars, at the end of which day will she have less than $10,000 left?

(A) 7
(B) 8
(C) 9
(D) 10
(E) 11

3. If $x < \frac{4}{z}$ and $z < 0$, then

(A) $x + z < 4$
(B) $x + z > 4$
(C) $\frac{x}{2} < 4$
(D) $xz < 4$
(E) $xz > 4$

GO ON

Name ______________________ Date __________ Class __________

SAT Practice Test 1 Section 3 *continued*

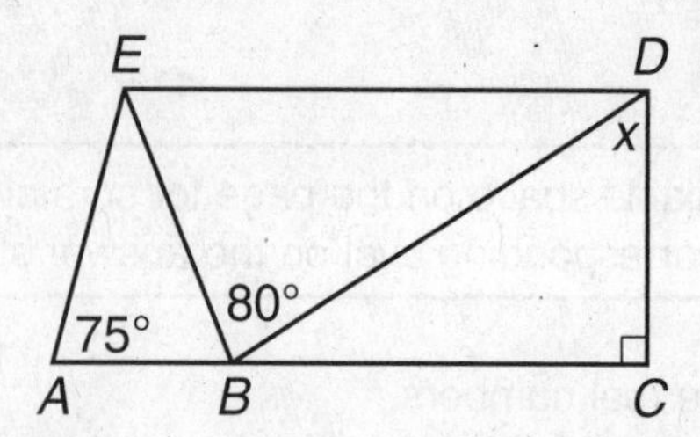

4. In the figure above $AE = BE$. What is the value of x in degrees?

(A) 65°

(B) 70°

(C) 75°

(D) 80°

(E) 85°

5. If $y = |x^2| - |x - 4| + |2x + 1|$ and $x = -3$, then $\frac{1}{y} =$

(A) $-\frac{1}{11}$

(B) $\frac{1}{21}$

(C) $\frac{1}{11}$

(D) $\frac{1}{9}$

(E) $\frac{1}{7}$

6. If p is a prime number and $p \neq 2$, which of the following is NOT an even number?

(A) $2p$

(B) $p + 1$

(C) $3p + 1$

(D) $3p - 2$

(E) $2p^2$

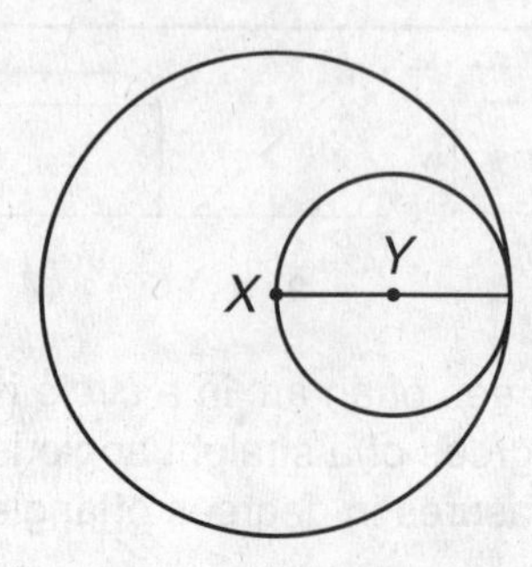

7. The diameter of $\odot X$ above is 40 inches. What is the ratio of the area of $\odot Y$ to the area of $\odot X$?

(A) $\frac{1}{16}$

(B) $\frac{1}{8}$

(C) $\frac{1}{4}$

(D) $\frac{1}{3}$

(E) $\frac{1}{2}$

8. If m is a positive odd integer and n is a negative even integer, what MUST be true about the sum of m and n?

(A) The sum is a negative even integer.

(B) The sum is a positive even integer.

(C) The sum is a negative odd integer.

(D) The sum is a positive odd integer.

(E) The sum cannot be equal to zero.

GO ON

9. Lines *m*, *n*, and *p* are all coplanar and there is at least one point on *m* that is not on *n* or *p*. If line *m* is perpendicular to line *n* and line *n* is perpendicular to line *p*, what MUST be true about the relationship between line *m* and line *p*?

(A) They are perpendicular lines.

(B) They are parallel lines.

(C) They are the same line.

(D) They intersect in exactly one point.

(E) Cannot be determined from the information given.

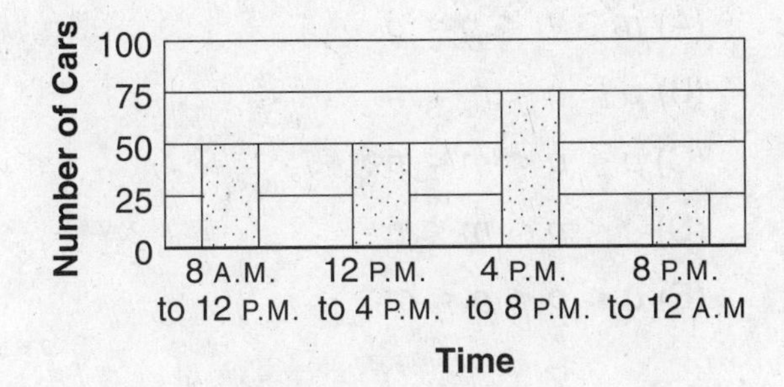

10. The number of cars entering a park yesterday between certain times is shown in the bar graph above. About what percent of the total number of cars entered the park after 4 P.M.?

(A) 0%

(B) 25%

(C) 50%

(D) 75%

(E) 100%

11. Point *A* is graphed on a coordinate grid. If the distance from the origin to point *A* is 5 units, which of the following could be the coordinates of *A*?

(A) (4, 3)

(B) (−4, 3)

(C) (−3, −4)

(D) (3, −4)

(E) All of the above.

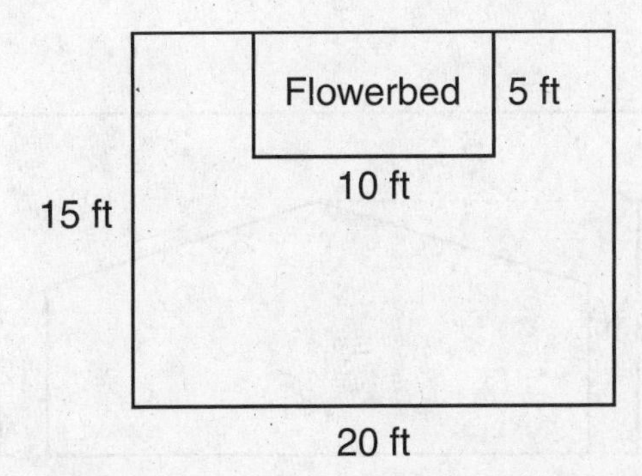

12. Jake's front yard is shown in the diagram above. If Eric throws a ball from down the street and it lands in Jake's front yard, what is the probability that the ball lands in the flowerbed?

(A) $\frac{1}{300}$

(B) $\frac{1}{50}$

(C) $\frac{1}{10}$

(D) $\frac{1}{6}$

(E) $\frac{1}{2}$

GO ON

13. Ramon can paint a fence in two hours. Jenny can paint the same fence 15 minutes faster than Ramon and Dennis can paint it in half Ramon's time. About how many minutes will it take to paint the fence if all three people work together?

(A) 22

(B) 29

(C) 42

(D) 48

(E) 60

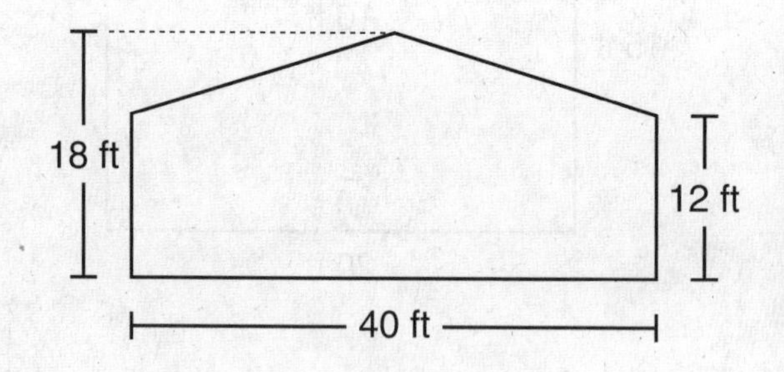

14. Ray is using a paint sprayer to paint the back of the barn shown above. If the sprayer can hold enough paint to cover 100 square feet, how many times will Ray have to fill the paint sprayer?

(A) 6

(B) 6.5

(C) 7

(D) 7.5

(E) 8

15. A set of data consists of four consecutive numbers. The sum of the numbers is 50. What is the median of the set of data?

(A) 12

(B) 12.5

(C) 13

(D) 13.5

(E) 14

16. If $m = 5$, $n = 2^3$, $p = 4^{-\frac{1}{2}}$, and $q = -0.25$, which of the following is true?

(A) $m < n < p < q$

(B) $p < q < n < m$

(C) $p < q < m < n$

(D) $q < p < m < n$

(E) $q < p < n < m$

STOP **If you finish before time is called, you may check your work on this section only. Do not turn to any other section in the test.**

Name ______________________ Date __________ Class __________

SAT Practice Test 2 Section 1

Time—25 minutes, 20 Questions

Directions: In this section, solve each problem using any available space on the page for scratch work. Then decide which is best of the choices given and fill in the corresponding oval on the answer sheet.

Notes:

1. The use of a calculator is permitted. All numbers used are real numbers.
2. Figures that accompany problems in this test are intended to provide information useful in solving the problems. They are drawn as accurately as possible EXCEPT when it is stated in a specific problem that the figure is not drawn to scale. All figures lie in a plane unless otherwise indicated.

Reference Information

$A = \pi r^2$
$C = 2\pi r$

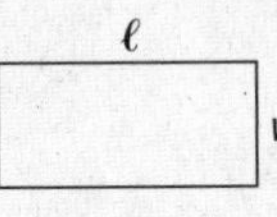

$A = \ell w$

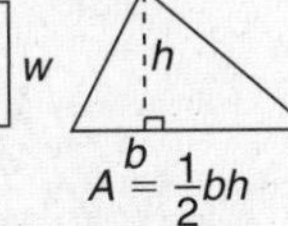

$A = \frac{1}{2}bh$

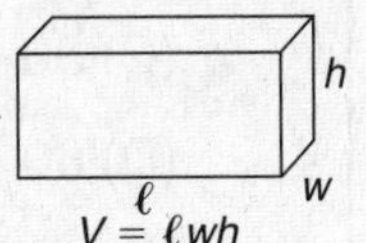

$V = \ell wh$

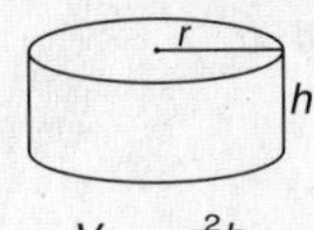

$V = \pi r^2 h$

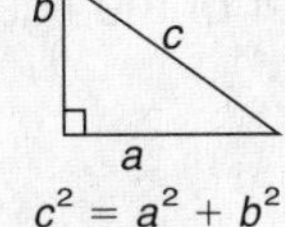

$c^2 = a^2 + b^2$

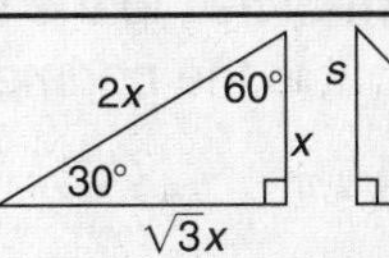

Special Right Triangles

The number of degrees of an arc in a circle is 360.
The measure in degrees of a straight angle is 180.
The sum of the measures in degrees of angles of a triangle is 180.

1. What is one-fourth of the sum of one-half and one-third?

(A) $\frac{1}{24}$

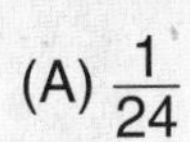

(B) $\frac{1}{12}$

(C) $\frac{1}{10}$

(D) $\frac{5}{24}$

(E) $\frac{2}{9}$

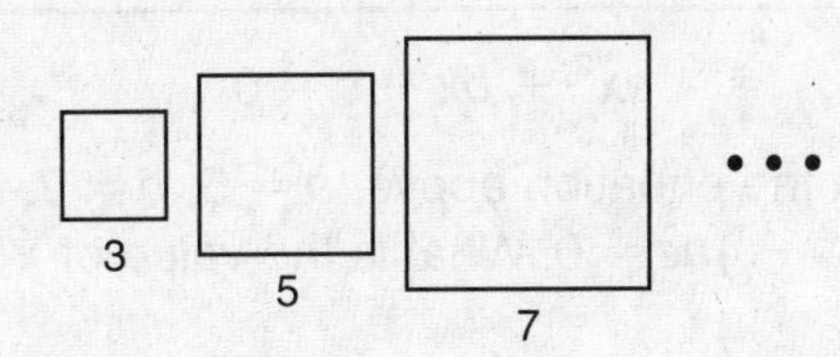

2. The figures above are squares. If the pattern continues, what would be the area of the tenth square?

(A) 6

(B) 91

(C) 196

(D) 361

(E) 441

3. If $y = \sqrt{x}$, then $2 + y^2 =$

(A) 2

(B) $2x$

(C) $2 + \sqrt{x}$

(D) $2 + x$

(E) $2 + x^2$

GO ON

Name ______________________ Date ____________ Class ____________

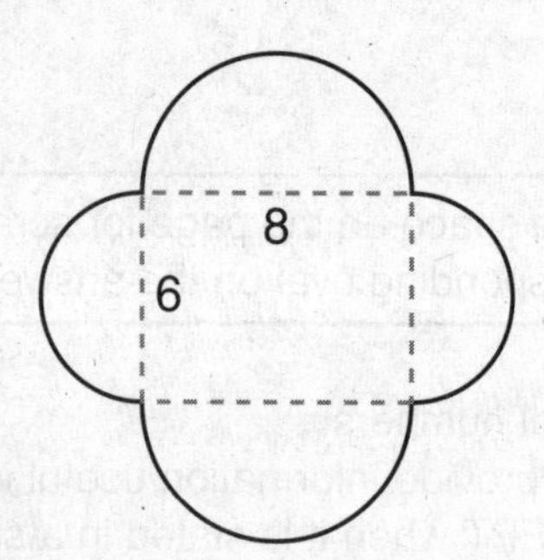

4. The figure above was made by connecting semicircles to the sides of a rectangle. What is the perimeter of the figure?

(A) 7π

(B) 14π

(C) $14\pi + 28$

(D) 28π

(E) $28\pi + 28$

$$ax^2 + bx + c = 0$$

5. In the equation above, $a = 5$, $b = 7$, and $b^2 - 4ac = 0$. What is the value of x?

(A) $-\frac{7}{5}$

(B) $-\frac{7}{10}$

(C) 0

(D) $\frac{7}{10}$

(E) $\frac{7}{5}$

---

6. If the second leg of a right triangle is three times as long as the first, then the hypotenuse is how many times as long as the first leg?

(A) $\sqrt{2}$

(B) 2

(C) $\sqrt{10}$

(D) 4

(E) 10

7. A local high school's tennis team has a tournament today. There are three matches in the tournament, *A*, *B*, and *C*, all of which are played at the same time. There are 5 players on the school team and only one player plays in each match. How many different ways can the coach assign the players?

(A) 12

(B) 15

(C) 30

(D) 60

(E) 120

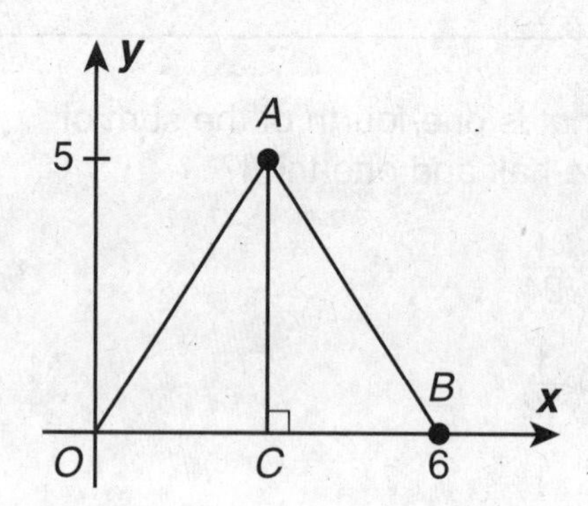

8. If *C* is the midpoint of $\overline{OB}$, what is an equation of the line that passes through $\overline{OA}$?

(A) $y = \frac{5}{6}x$

(B) $y = \frac{5}{6}x + 5$

(C) $y = \frac{5}{6}x + 6$

(D) $y = \frac{5}{3}x$

(E) $y = \frac{5}{3}x + 5$

GO ON

Name ______________________ Date __________ Class __________

9. Mr. Lawrence stops at a gas station. He has enough cash in his pocket to buy nine gallons of gas. If the cost of each gallon of gas was 20 cents less, Mr. Lawrence could buy one more gallon. How much cash does Mr. Lawrence have?

(A) \$18.00

(B) \$19.00

(C) \$20.00

(D) \$22.00

(E) \$25.00

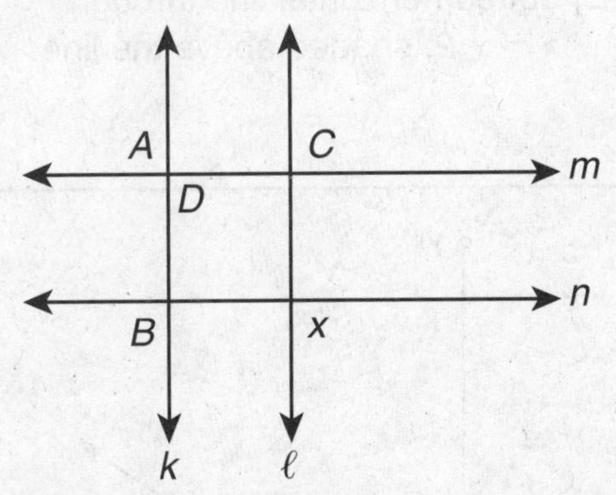

10. If line m is parallel to line n, line k is parallel to line ℓ, and $x = 90°$, which of the following does NOT equal x?

(A) $\angle A$

(B) $\angle B$

(C) $\angle C$

(D) $\angle D$

(E) None of the above.

11. If $2n^2$ is an even integer, what is true about $n + 1$?

(A) It is an even integer.

(B) It is an odd integer.

(C) It can be an odd or even integer.

(D) It is a prime number.

(E) It is a perfect square.

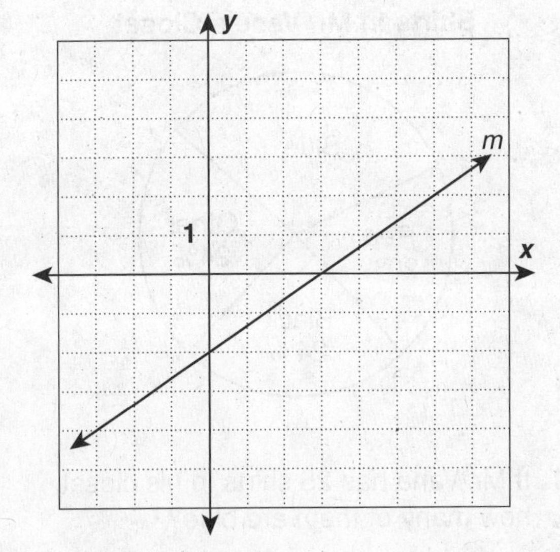

12. What is the slope of line n if line $n \perp$ line m?

(A) $-\frac{3}{2}$

(B) $-\frac{2}{3}$

(C) $\frac{2}{3}$

(D) $\frac{3}{2}$

(E) Cannot be determined from the information provided.

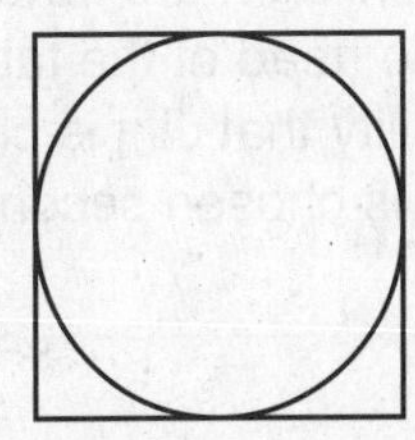

13. The circle shown is inscribed inside a square of area 36 units^2. What is the area of the circle in square units?

(A) 6π

(B) 9π

(C) 20.25π

(D) 20.5π

(E) 36π

GO ON

Shirts in Mr. Vane's Closet

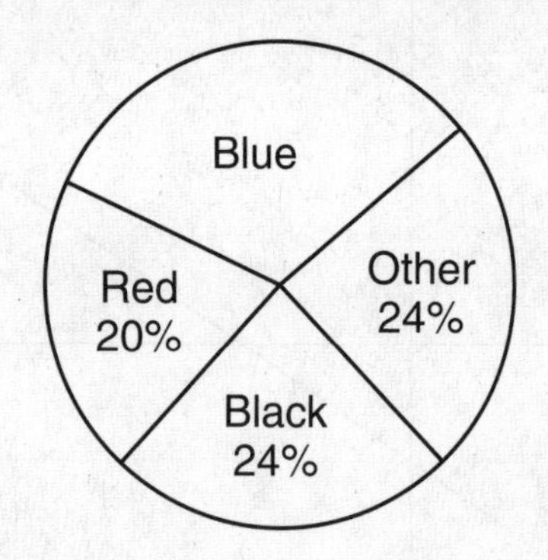

14. If Mr. Vane has 25 shirts in his closet, how many of them are blue?

(A) 4

(B) 6

(C) 8

(D) 25

(E) 32

---

15. There are 10 children in Jim's family including Jim and his brother Aaron. If two of the children are randomly chosen to eat at the head of the table, what is the probability that Jim is chosen first and Aaron is chosen second?

(A) $\frac{1}{100}$

(B) $\frac{1}{90}$

(C) $\frac{1}{19}$

(D) $\frac{1}{10}$

(E) $\frac{1}{5}$

$$2x + 3 < 3x + 5$$

16. If the solution set of the inequality above were graphed on a set of coordinate axes, the graph would be a

(A) solid vertical line through $x = -2$, shaded to the left.

(B) solid vertical line through $x = -2$, shaded to the right.

(C) dotted vertical line through $x = -2$, shaded to the left.

(D) dotted vertical line through $x = -2$, shaded to the right.

(E) dotted horizontal line through $x = -2$, shaded above the line.

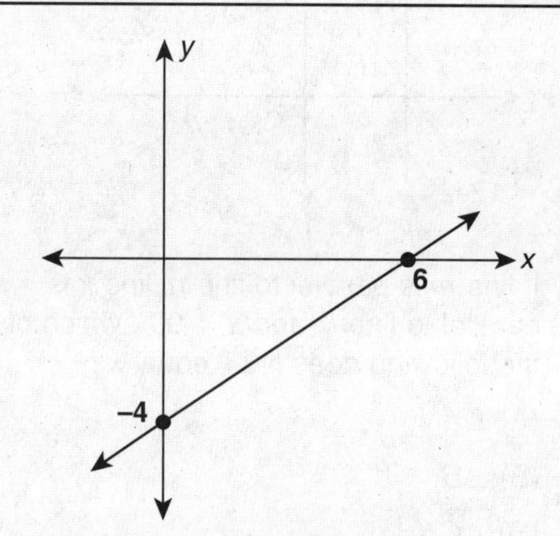

17. If the line shown above is reflected over the x-axis, what is the equation of the new line?

(A) $y = -\frac{2}{3}x + 4$

(B) $y = \frac{2}{3}x - 4$

(C) $y = -\frac{2}{3}x - 4$

(D) $y = \frac{2}{3}x + 4$

(E) $y = -\frac{3}{2}x + 4$

GO ON

$$x - 3y \leq 4$$
$$3x + y \leq 4$$

18. Which point satisfies the system of linear inequalities?

(A) $(-1, -2)$

(B) $(-1, 2)$

(C) $(1, 2)$

(D) $(2, -3)$

(E) $(2, 3)$

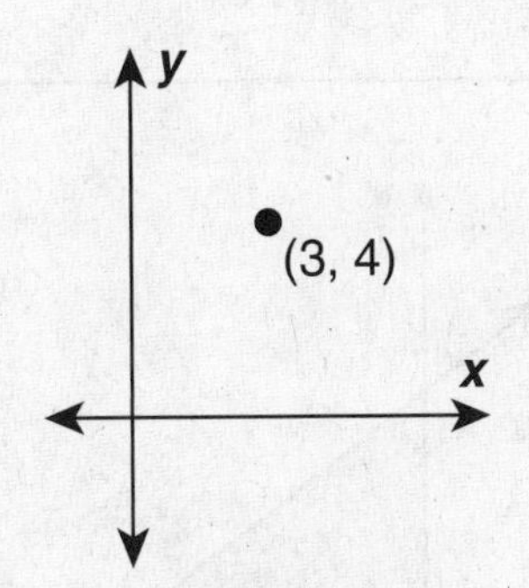

19. A square is to be drawn on the coordinate axes shown above. Its bottom, left-hand corner is to be the point (3, 4) and its opposite sides are to be parallel to the axes. Which of the following could be the coordinates of the square's upper, right-hand corner?

(A) (0, 0)

(B) (4, 3)

(C) (7, 6)

(D) (8, 6)

(E) (8, 9)

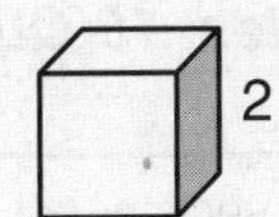

20. The figure shown above is a cube. If the sides of the cube are increased by 3 units, by how many cubic units would the volume increase?

(A) 3

(B) 9

(C) 27

(D) 81

(E) 117

STOP **If you finish before time is called, you may check your work on this section only. Do not turn to any other section in the test.**

Name ______________________ Date __________ Class __________

SAT Practice Test 2 Section 2

Time—25 minutes, 18 Questions

Directions: In this section, solve each problem using any available space on the page for scratch work. Then decide which is best of the choices given and fill in the corresponding oval on the answer sheet.

Notes:

1. The use of a calculator is permitted. All numbers used are real numbers.
2. Figures that accompany problems in this test are intended to provide information useful in solving the problems. They are drawn as accurately as possible EXCEPT when it is stated in a specific problem that the figure is not drawn to scale. All figures lie in a plane unless otherwise indicated.

Reference Information

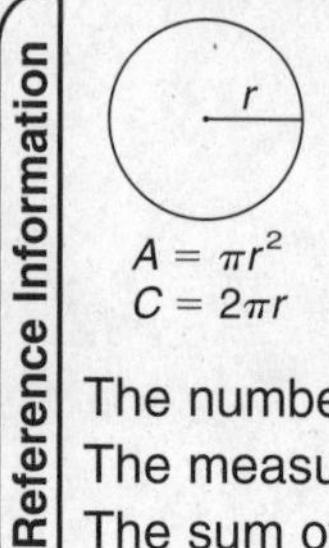

$A = \pi r^2$
$C = 2\pi r$

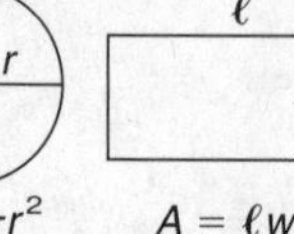

$A = \ell w$

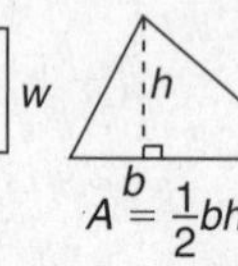

$A = \frac{1}{2}bh$

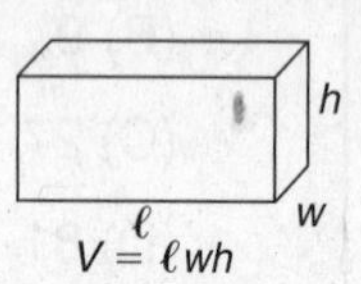

$V = \ell wh$

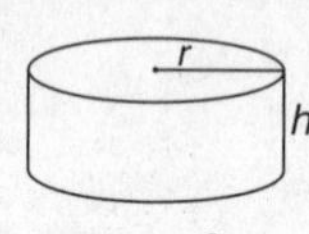

$V = \pi r^2 h$

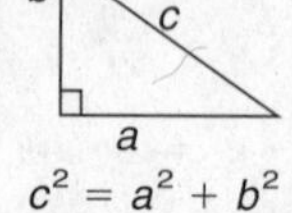

$c^2 = a^2 + b^2$

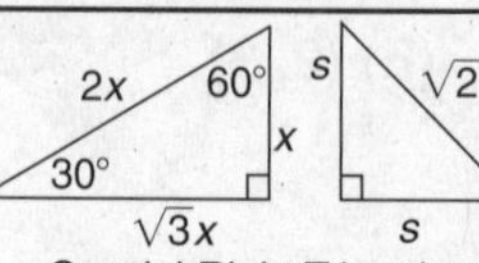

Special Right Triangles

The number of degrees of an arc in a circle is 360.
The measure in degrees of a straight angle is 180.
The sum of the measures in degrees of angles of a triangle is 180.

1. There are less than 50 coins in a jar. There are 17 pennies, 20 dimes and 2 quarters. If the rest of the coins are nickels, which of the following could be the value of the coins in the jar?

(A) \$2.62

(B) \$2.76

(C) \$2.90

(D) \$3.07

(E) \$3.32

2. If $5^{3x} \cdot 25^{x+1} = 5^{x+2}$, then $x =$

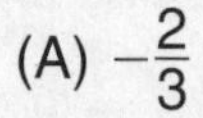

(A) $-\frac{2}{3}$

(B) 0

(C) $\frac{1}{3}$

(D) $\frac{3}{2}$

(E) 5

y
x
m

3. The equation of line m is $y = -\frac{3}{4}x + 6$. What is the area of the shaded region in square units?

(A) 12

(B) 24

(C) 36

(D) 48

(E) 50

4. If $\frac{\frac{1}{x}}{1 + \frac{1}{x}} = 3$, what does $\frac{1}{x}$ equal?

(A) -3

(B) $-\frac{3}{2}$

(C) 0

(D) $\frac{3}{2}$

(E) 3

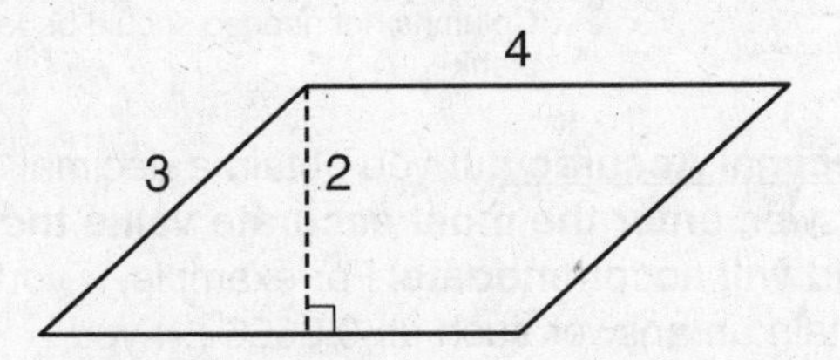

5. What would be the radius of a circle that has the same area as the parallelogram shown above?

(A) $2\sqrt{\pi}$

(B) $2\sqrt{2\pi}$

(C) $\frac{\sqrt{2\pi}}{\pi}$

(D) $\frac{2\sqrt{\pi}}{\pi}$

(E) $\frac{2\sqrt{2\pi}}{\pi}$

6. Joe is given five tests that all have a maximum score of 100 points. The mean of his test scores is 90. If all his tests have a different grade and all the grades are whole numbers, what is the lowest grade he could have scored?

(A) 44

(B) 48

(C) 56

(D) 58

(E) 62

7. Leroy travels often for his job. He has figured that he must pay \$650 per month for rent, \$2 per day for water for every day he is in his apartment, and \$4 per day for electricity for every day he is in his apartment. If x represents the number of days Leroy is in his apartment, which of the following represents the amount Leroy pays each month for rent, water, and electricity?

(A) $m(x) = x + 656$

(B) $m(x) = x + 830$

(C) $m(x) = 6x + 650$

(D) $m(x) = 650x$

(E) $m(x) = 830x$

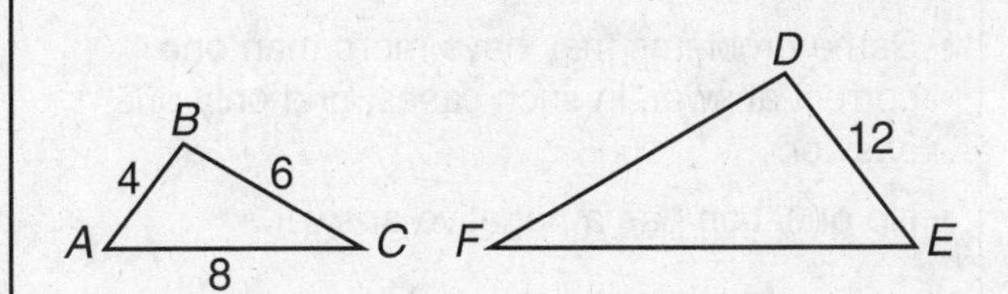

8. If $\triangle ABC \sim \triangle DEF$, which of the following statements is true? [Figures are not drawn to scale.]

(A) $DF = 8$

(B) $DF = 16$

(C) $EF = 8$

(D) $EF = 16$

(E) $EF = 18$

GO ON

SAT Practice Test 2 Section 2 *continued*

Directions for Student Response Questions

Each of the remaining 10 questions (11-20) require you to solve the problem and enter your answer by marking the ovals in the special grid, as shown in the examples below.

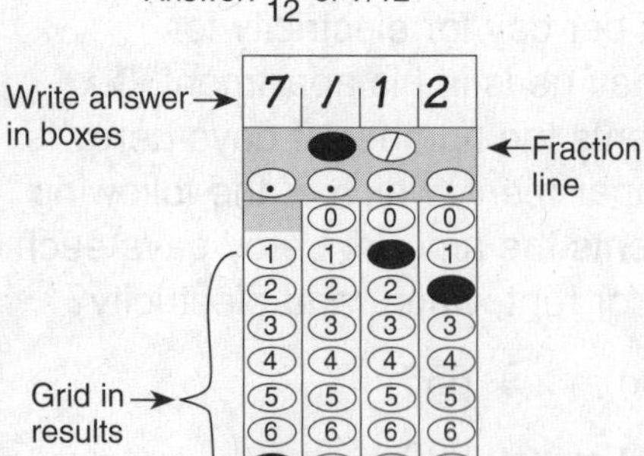

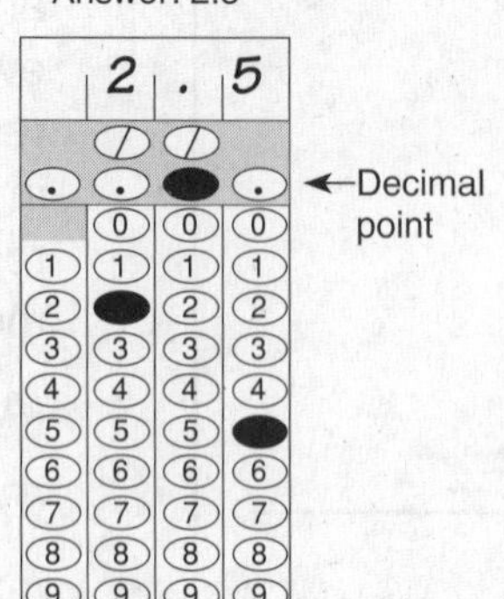

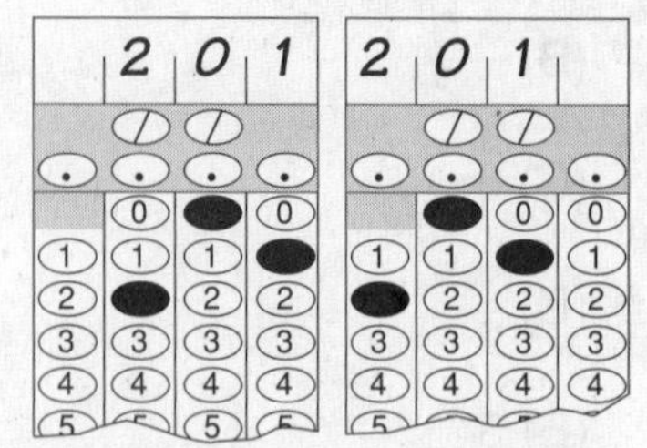

Note: You may start your answers in any column, space permitting. Columns not needed should be left blank

- Mark no more than one oval in any column.
- Because the answer sheet will be machined scored, **you will receive credit only if the ovals are filled in correctly**.
- Although not required, it is suggested that you write your answers in the boxes at the top of the columns to help you fill in the ovals accurately.
- Some problems may have more than one correct answer. In such cases, grid only one answer.
- No question has a negative answer.
- **Mixed numbers** such as $2\frac{1}{2}$ must be gridded as 2.5 or 5/2. (If 2 1 / 2 is gridded, it will be interpreted as $\frac{21}{2}$, not $2\frac{1}{2}$.)
- **Decimal Accuracy**: If you obtain a decimal answer, **enter the most accurate value the grid will accommodate**. For example, if you obtain an answer such as 0.6666 …, you should record the result as .666 or .667. **Less accurate values such as .66 or .67 are not acceptable.**

Acceptable ways to grid $\frac{2}{3}$ = .6666 …

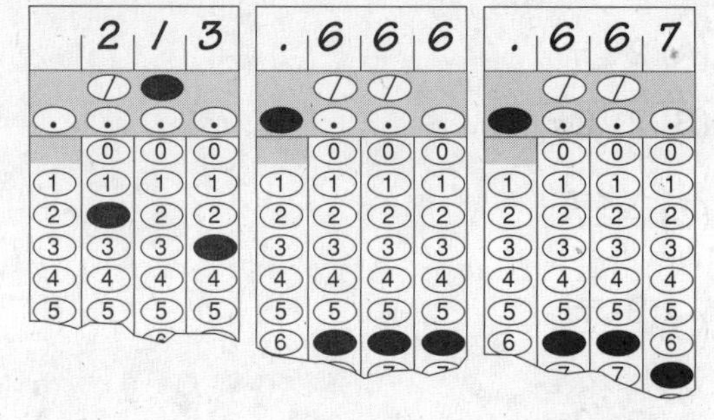

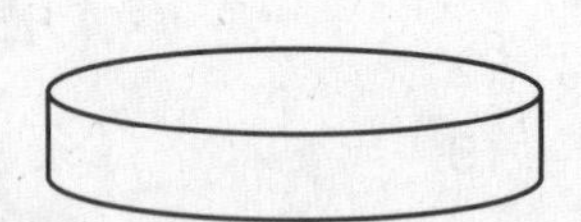

9. The radius of the cylinder shown above is three times the height. If its volume is 576π cubic units, what is the radius?

10. Mary, Fred, and Jane are all different ages. If the product of their ages is 27, what is the sum of their ages? [Ages are in whole numbers.]

GO ON

SAT Practice Test 2 Section 2 *continued*

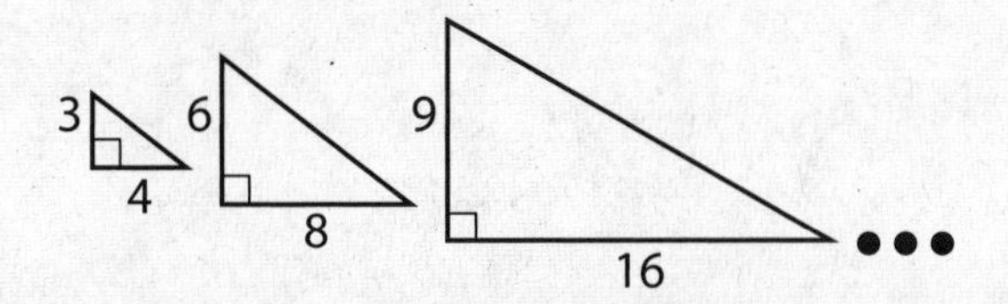

11. If the sequence above continues, what is the area of the sixth triangle in square units?

---

12. The variables x and y have an inverse variation. When $x = 2$, $y = 5$. What is the sum of x and y when $x = 1$?

---

13. What is the slope of every line that is parallel to $x = \frac{3}{2}y - 5$?

14. If x is an integer and x, $x + 2$, and $x + 4$ form a Pythagorean triple, what is the value of x?

$$2x + 3y = 12$$
$$-5x + 7y = -14$$

15. What is the distance between the y-intercepts of the graphs of the given equations?

---

16. If the roots of $2x^2 + bx + 9 = 0$ are $x = -\frac{1}{2}$ and $x = -9$, what is the value of b?

GO ON

Name ______________________ Date __________ Class __________

SAT Practice Test 2 Section 2 *continued*

(Exercises 17 and 18)

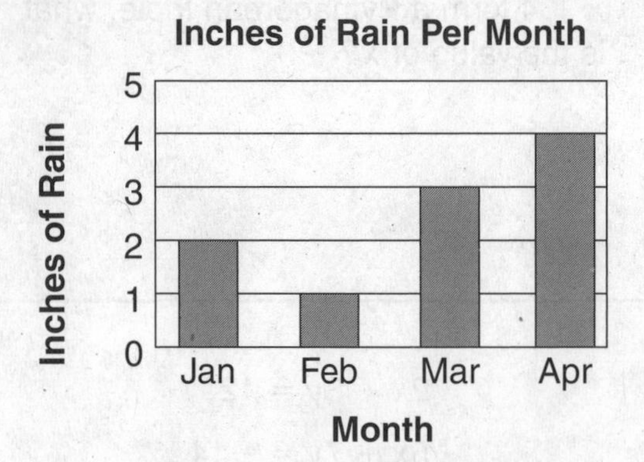

17. According to the graph above, what was the percent increase in rainfall from February to March?

---

18. What is the mean amount of rain, in inches, that fell during the first four months of the year?

STOP **If you finish before time is called, you may check your work on this section only. Do not turn to any other section in the test.**

Name ______________________________ Date ____________ Class ____________

SAT Practice Test 2 Section 3

Time—20 minutes, 16 Questions

Directions: In this section, solve each problem using any available space on the page for scratch work. Then decide which is best of the choices given and fill in the corresponding oval on the answer sheet.

Notes:

1. The use of a calculator is permitted. All numbers used are real numbers.
2. Figures that accompany problems in this test are intended to provide information useful in solving the problems. They are drawn as accurately as possible EXCEPT when it is stated in a specific problem that the figure is not drawn to scale. All figures lie in a plane unless otherwise indicated.

Reference Information

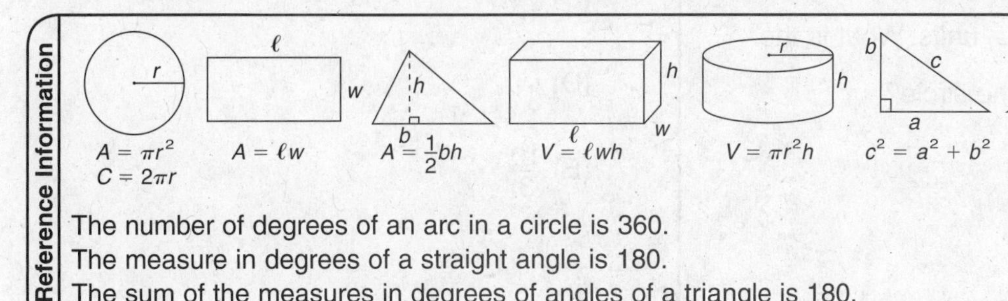

The number of degrees of an arc in a circle is 360.
The measure in degrees of a straight angle is 180.
The sum of the measures in degrees of angles of a triangle is 180.

1. Jared has a collection of 90 movies on DVD. If 46 of the movies are comedies, 26 are drama, and the rest are science fiction, what percent of the movies are science fiction?

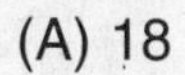

(A) 18

(B) 20

(C) 28

(D) 32

(E) 35

2. If $m \Psi n = n + mn$, what is $2\Psi(3 \Psi 4)$?

(A) 32

(B) 34

(C) 40

(D) 48

(E) 50

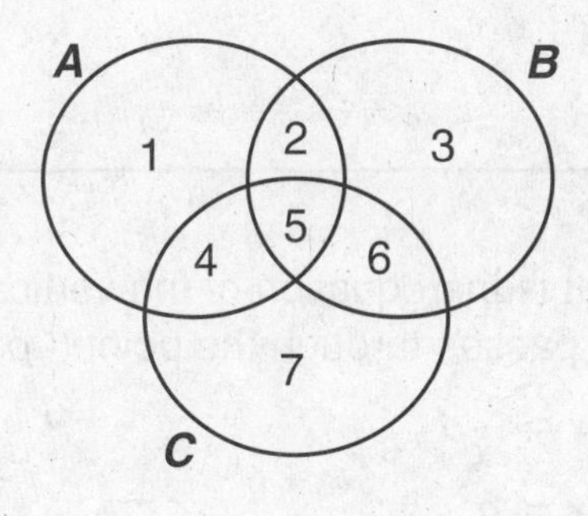

3. What is the sum of the elements of $B \cap C$?

(A) 9

(B) 11

(C) 15

(D) 27

(E) 38

GO ON

Name ______________________ Date __________ Class __________

SAT Practice Test 2 Section 3 *continued*

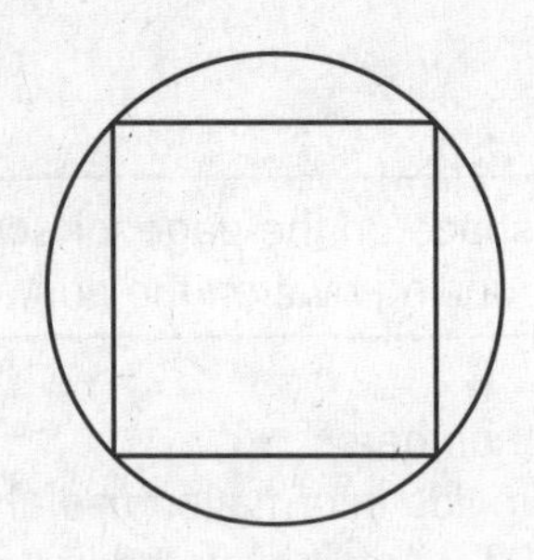

4. The length of the sides of the square shown above is $\frac{\sqrt{2}}{\pi}$ units. What is the circumference of the circle?

(A) 2

(B) 2π

(C) 4

(D) 4π

(E) 8

---

5. What is the equation of the vertical line that passes through the point $(-p, q)$?

(A) $x = -p$

(B) $x = p$

(C) $y = -q$

(D) $y = p$

(E) $y = -\frac{q}{p}$

---

6. If $a = b^2$, $b = c^2$, and $a = 64$, what is the value of c^3?

(A) 4

(B) 8

(C) $8\sqrt{2}$

(D) 16

(E) $16\sqrt{2}$

7. There are 20 videos on a shelf. Lee likes six of the videos. If Chris chooses one video at random, what is the probability that he picks one that Lee does not like?

(A) $\frac{3}{10}$

(B) $\frac{3}{7}$

(C) $\frac{7}{10}$

(D) $\frac{7}{3}$

(E) $\frac{10}{3}$

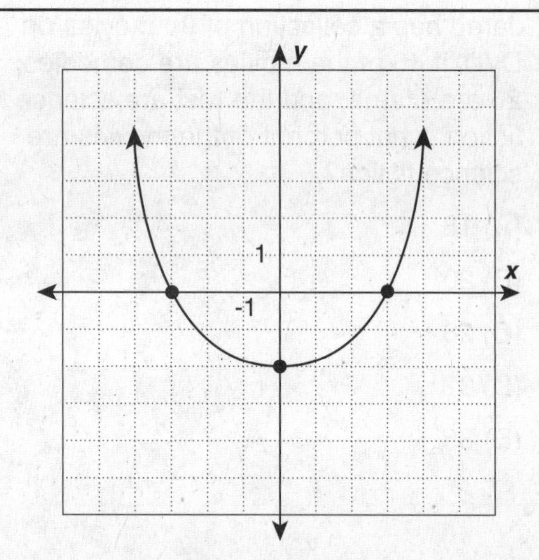

8. The graph of $f(x)$ is shown above. What is the value of $|f(0)|$?

(A) -3

(B) -2

(C) 0

(D) 2

(E) 3

GO ON

SAT Practice Test 2 Section 3 *continued*

9. If $x = t + 1$ and $y = x^2 - x - 1$, which of the following is equal to y?

(A) $t^2 - t - 1$

(B) $t^2 - t + 1$

(C) $t^2 + t - 1$

(D) $t^2 + t + 1$

(E) $t^2 + t + 2$

$$x^2 + y^2 = 4$$
$$2x + y = 4$$

10. What is the positive difference between the x-coordinates of the two points of intersection between the two graphs?

(A) $\frac{4}{5}$

(B) 1

(C) $\frac{6}{5}$

(D) $\frac{7}{5}$

(E) $\frac{8}{5}$

---

11. The mean of six data values is 18. If all the values are different positive whole numbers, what is the greatest value that the range could be?

(A) 90

(B) 91

(C) 92

(D) 93

(E) 94

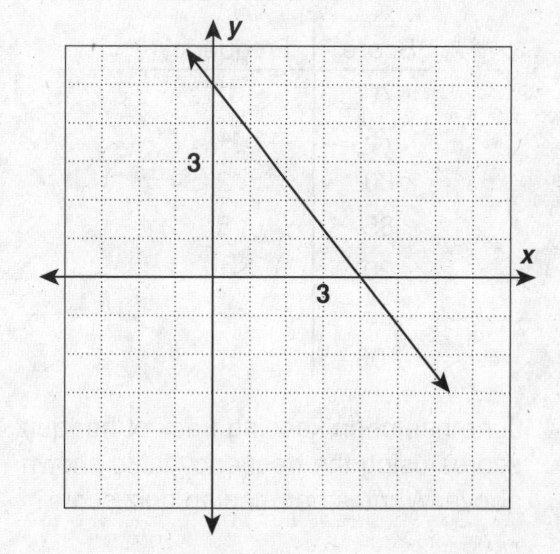

12. If the line above is shifted down 2 units and to the right 4 units, what is the equation of the new line?

(A) $y = -\frac{5}{4}x + 2$

(B) $y = -\frac{5}{4}x + 5$

(C) $y = -\frac{5}{4}x + 8$

(D) $y = -\frac{1}{2}x + 2$

(E) $y = -\frac{1}{2}x + 5$

---

13. If $\sqrt{-3x + \sqrt{15 - x}} = \sqrt{x}$, what is the value of x?

(A) $\frac{1}{3}$

(B) $\frac{5}{8}$

(C) $\frac{3}{4}$

(D) $\frac{15}{16}$

(E) 1

GO ON

SAT Practice Test 2 Section 3 *continued*

Score	Frequency
70	12
75	4
80	2
85	6
90	1
95	2
100	1

14. June has been keeping track of her quiz scores using the frequency table shown above. What is her median quiz score?

(A) 70

(B) 75

(C) 80

(D) 85

(E) 90

$$\frac{3}{10} + \frac{4}{10^3} + \frac{5}{10^5} + \ldots$$

15. What is the sum of the first six terms of the series above?

(A) 0.0304050607

(B) 0.030405060708

(C) 0.30405060708

(D) 0.3456

(E) 0.345678

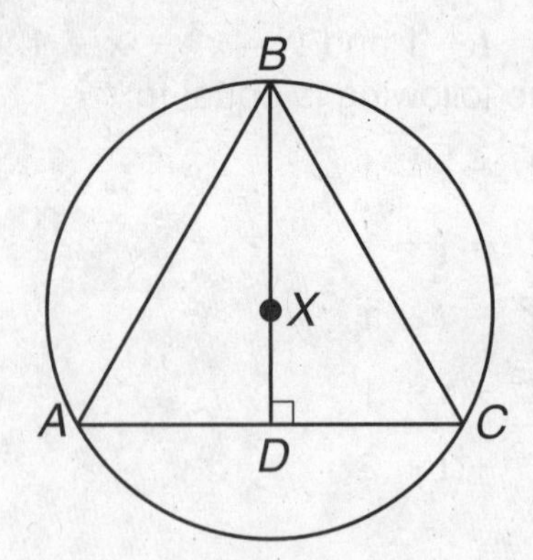

16. Point X is the center of the circle shown above. If the radius of the circle is 2 and $\triangle ABC$ is an equilateral triangle, what is the length of $\overline{XD}$?

(A) $\frac{\sqrt{2}}{3}$

(B) $\frac{\sqrt{3}}{2}$

(C) 1

(D) $\sqrt{2}$

(E) $\sqrt{3}$

STOP If you finish before time is called, you may check your work on this section only. Do not turn to any other section in the test.

Name ____________________ Date __________ Class __________

SAT Subject Test Practice Test I: Math Level IC

1. YOUR NAME: ______________________________
(Print) Last First M.I.

SIGNATURE: ____________________ **DATE:** __________

HOME ADDRESS: ______________________________
(Print) Number and Street

______________________________ **E-MAIL:** __________
City State Zip

PHONE NO.: __________ **SCHOOL:** __________ **CLASS OF:** __________
(Print)

Completely darken bubbles with a No. 2 pencil. If you make a mistake, be sure to erase mark completely. Erase all stray marks.

IMPORTANT: Please fill in these boxes exactly as shown on the back cover of your text book.

2. TEST FORM

3. TEST CODE

0	A	J	0	0
1	B	K	1	1
2	C	L	2	2
3	D	M	3	3
4	E	N	4	4
5	F	O	5	5
6	G	P	6	6
7	H	Q	7	7
8	I	R	8	8
9			9	9

4. REGISTRATION NUMBER

0	0	0	0	0	0
1	1	1	1	1	1
2	2	2	2	2	2
3	3	3	3	3	3
4	4	4	4	4	4
5	5	5	5	5	5
6	6	6	6	6	6
7	7	7	7	7	7
8	8	8	8	8	8
9	9	9	9	9	9

5. YOUR NAME

First 4 letters of last name				FIRST INT	LAST INT
A–Z	A–Z	A–Z	A–Z	A–Z	A–Z

6. DATE OF BIRTH

MONTH	DAY		YEAR	
JAN	0	0	0	0
FEB	1	1	1	1
MAR	2	2	2	2
APR	3	3	3	3
MAY		4	4	4
JUN		5	5	5
JUL		6	6	6
AUG		7	7	7
SEP		8	8	8
OCT		9	9	9
NOV				
DEC				

7. SEX

MALE
FEMALE

1	A B C D E	11	A B C D E	21	A B C D E	31	A B C D E	41	A B C D E
2	A B C D E	12	A B C D E	22	A B C D E	32	A B C D E	42	A B C D E
3	A B C D E	13	A B C D E	23	A B C D E	33	A B C D E	43	A B C D E
4	A B C D E	14	A B C D E	24	A B C D E	34	A B C D E	44	A B C D E
5	A B C D E	15	A B C D E	25	A B C D E	35	A B C D E	45	A B C D E
6	A B C D E	16	A B C D E	26	A B C D E	36	A B C D E	46	A B C D E
7	A B C D E	17	A B C D E	27	A B C D E	37	A B C D E	47	A B C D E
8	A B C D E	18	A B C D E	28	A B C D E	38	A B C D E	48	A B C D E
9	A B C D E	19	A B C D E	29	A B C D E	39	A B C D E	49	A B C D E
10	A B C D E	20	A B C D E	30	A B C D E	40	A B C D E	50	A B C D E

Name ______________________ Date __________ Class __________

SAT Subject Test Practice Test II: Math Level IC

1. YOUR NAME: (Print) ______ Last ______ First ______ M.I.

SIGNATURE: ______ **DATE:** ______

HOME ADDRESS: (Print) ______ Number and Street

______ City ______ State ______ Zip **E-MAIL:** ______

PHONE NO.: (Print) ______ **SCHOOL:** ______ **CLASS OF:** ______

Completely darken bubbles with a No. 2 pencil. If you make a mistake, be sure to erase mark completely. Erase all stray marks.

IMPORTANT: Please fill in these boxes exactly as shown on the back cover of your text book.

2. TEST FORM

3. TEST CODE

0	A	J	0	0
1	B	K	1	1
2	C	L	2	2
3	D	M	3	3
4	E	N	4	4
5	F	O	5	5
6	G	P	6	6
7	H	Q	7	7
8	I	R	8	8
9			9	9

4. REGISTRATION NUMBER

0	0	0	0	0	0
1	1	1	1	1	1
2	2	2	2	2	2
3	3	3	3	3	3
4	4	4	4	4	4
5	5	5	5	5	5
6	6	6	6	6	6
7	7	7	7	7	7
8	8	8	8	8	8
9	9	9	9	9	9

5. YOUR NAME

First 4 letters of last name				FIRST INT	LAST INT
A	A	A	A	A	A
B	B	B	B	B	B
C	C	C	C	C	C
D	D	D	D	D	D
E	E	E	E	E	E
F	F	F	F	F	F
G	G	G	G	G	G
H	H	H	H	H	H
I	I	I	I	I	I
J	J	J	J	J	J
K	K	K	K	K	K
L	L	L	L	L	L
M	M	M	M	M	M
N	N	N	N	N	N
O	O	O	O	O	O
P	P	P	P	P	P
Q	Q	Q	Q	Q	Q
R	R	R	R	R	R
S	S	S	S	S	S
T	T	T	T	T	T
U	U	U	U	U	U
V	V	V	V	V	V
W	W	W	W	W	W
X	X	X	X	X	X
Y	Y	Y	Y	Y	Y
Z	Z	Z	Z	Z	Z

6. DATE OF BIRTH

MONTH	DAY		YEAR	
JAN	0	0	0	0
FEB	1	1	1	1
MAR	2	2	2	2
APR	3	3	3	3
MAY		4	4	4
JUN		5	5	5
JUL		6	6	6
AUG		7	7	7
SEP		8	8	8
OCT		9	9	9
NOV				
DEC				

7. SEX

- MALE
- FEMALE

1	A B C D E	11	A B C D E	21	A B C D E	31	A B C D E	41	A B C D E
2	A B C D E	12	A B C D E	22	A B C D E	32	A B C D E	42	A B C D E
3	A B C D E	13	A B C D E	23	A B C D E	33	A B C D E	43	A B C D E
4	A B C D E	14	A B C D E	24	A B C D E	34	A B C D E	44	A B C D E
5	A B C D E	15	A B C D E	25	A B C D E	35	A B C D E	45	A B C D E
6	A B C D E	16	A B C D E	26	A B C D E	36	A B C D E	46	A B C D E
7	A B C D E	17	A B C D E	27	A B C D E	37	A B C D E	47	A B C D E
8	A B C D E	18	A B C D E	28	A B C D E	38	A B C D E	48	A B C D E
9	A B C D E	19	A B C D E	29	A B C D E	39	A B C D E	49	A B C D E
10	A B C D E	20	A B C D E	30	A B C D E	40	A B C D E	50	A B C D E

Name ______________________ Date __________ Class __________

SAT Subject Test Practice Test I: Math Level IIC

1. YOUR NAME: ______________________________
(Print) Last First M.I.

SIGNATURE: ______________________ **DATE:** __________

HOME ADDRESS: ______________________________
(Print) Number and Street

______________________ **E-MAIL:** __________
City State Zip

PHONE NO.: __________ **SCHOOL:** __________ **CLASS OF:** __________
(Print)

Completely darken bubbles with a No. 2 pencil. If you make a mistake, be sure to erase mark completely. Erase all stray marks.

IMPORTANT: Please fill in these boxes exactly as shown on the back cover of your text book.

2. TEST FORM

3. TEST CODE

0	A	J	0	0
1	B	K	1	1
2	C	L	2	2
3	D	M	3	3
4	E	N	4	4
5	F	O	5	5
6	G	P	6	6
7	H	Q	7	7
8	I	R	8	8
9			9	9

4. REGISTRATION NUMBER

0	0	0	0	0	0
1	1	1	1	1	1
2	2	2	2	2	2
3	3	3	3	3	3
4	4	4	4	4	4
5	5	5	5	5	5
6	6	6	6	6	6
7	7	7	7	7	7
8	8	8	8	8	8
9	9	9	9	9	9

5. YOUR NAME

First 4 letters of last name				FIRST INT	LAST INT
A	A	A	A	A	A
B	B	B	B	B	B
C	C	C	C	C	C
D	D	D	D	D	D
E	E	E	E	E	E
F	F	F	F	F	F
G	G	G	G	G	G
H	H	H	H	H	H
I	I	I	I	I	I
J	J	J	J	J	J
K	K	K	K	K	K
L	L	L	L	L	L
M	M	M	M	M	M
N	N	N	N	N	N
O	O	O	O	O	O
P	P	P	P	P	P
Q	Q	Q	Q	Q	Q
R	R	R	R	R	R
S	S	S	S	S	S
T	T	T	T	T	T
U	U	U	U	U	U
V	V	V	V	V	V
W	W	W	W	W	W
X	X	X	X	X	X
Y	Y	Y	Y	Y	Y
Z	Z	Z	Z	Z	Z

6. DATE OF BIRTH

MONTH	DAY		YEAR	
JAN				
FEB				
MAR	0	0	0	0
APR	1	1	1	1
MAY	2	2	2	2
JUN	3	3	3	3
JUL		4	4	4
AUG		5	5	5
SEP		6	6	6
OCT		7	7	7
NOV		8	8	8
DEC		9	9	9

7. SEX

- MALE
- FEMALE

1	A B C D E	11	A B C D E	21	A B C D E	31	A B C D E	41	A B C D E
2	A B C D E	12	A B C D E	22	A B C D E	32	A B C D E	42	A B C D E
3	A B C D E	13	A B C D E	23	A B C D E	33	A B C D E	43	A B C D E
4	A B C D E	14	A B C D E	24	A B C D E	34	A B C D E	44	A B C D E
5	A B C D E	15	A B C D E	25	A B C D E	35	A B C D E	45	A B C D E
6	A B C D E	16	A B C D E	26	A B C D E	36	A B C D E	46	A B C D E
7	A B C D E	17	A B C D E	27	A B C D E	37	A B C D E	47	A B C D E
8	A B C D E	18	A B C D E	28	A B C D E	38	A B C D E	48	A B C D E
9	A B C D E	19	A B C D E	29	A B C D E	39	A B C D E	49	A B C D E
10	A B C D E	20	A B C D E	30	A B C D E	40	A B C D E	50	A B C D E

Name ______________________ Date ____________ Class ____________

SAT Subject Test Practice Test II: Math Level IIC

1. YOUR NAME: ______________________
(Print) Last First M.I.

SIGNATURE: ______________________ **DATE:** ____________

HOME ADDRESS: ______________________
(Print) Number and Street

______________________ **E-MAIL:** ____________
City State Zip

PHONE NO.: ____________ **SCHOOL:** ____________ **CLASS OF:** ____________
(Print)

Completely darken bubbles with a No. 2 pencil. If you make a mistake, be sure to erase mark completely. Erase all stray marks.

IMPORTANT: Please fill in these boxes exactly as shown on the back cover of your text book.

2. TEST FORM

3. TEST CODE

0	A	J	0	0
1	B	K	1	1
2	C	L	2	2
3	D	M	3	3
4	E	N	4	4
5	F	O	5	5
6	G	P	6	6
7	H	Q	7	7
8	I	R	8	8
9			9	9

4. REGISTRATION NUMBER

0	0	0	0	0	0
1	1	1	1	1	1
2	2	2	2	2	2
3	3	3	3	3	3
4	4	4	4	4	4
5	5	5	5	5	5
6	6	6	6	6	6
7	7	7	7	7	7
8	8	8	8	8	8
9	9	9	9	9	9

5. YOUR NAME

First 4 letters of last name				FIRST INT	LAST INT
A	A	A	A	A	A
B	B	B	B	B	B
C	C	C	C	C	C
D	D	D	D	D	D
E	E	E	E	E	E
F	F	F	F	F	F
G	G	G	G	G	G
H	H	H	H	H	H
I	I	I	I	I	I
J	J	J	J	J	J
K	K	K	K	K	K
L	L	L	L	L	L
M	M	M	M	M	M
N	N	N	N	N	N
O	O	O	O	O	O
P	P	P	P	P	P
Q	Q	Q	Q	Q	Q
R	R	R	R	R	R
S	S	S	S	S	S
T	T	T	T	T	T
U	U	U	U	U	U
V	V	V	V	V	V
W	W	W	W	W	W
X	X	X	X	X	X
Y	Y	Y	Y	Y	Y
Z	Z	Z	Z	Z	Z

6. DATE OF BIRTH

MONTH	DAY		YEAR	
JAN				
FEB				
MAR	0	0	0	0
APR	1	1	1	1
MAY	2	2	2	2
JUN	3	3	3	3
JUL		4	4	4
AUG		5	5	5
SEP		6	6	6
OCT		7	7	7
NOV		8	8	8
DEC		9	9	9

7. SEX

MALE

FEMALE

1	A B C D E	11	A B C D E	21	A B C D E	31	A B C D E	41	A B C D E
2	A B C D E	12	A B C D E	22	A B C D E	32	A B C D E	42	A B C D E
3	A B C D E	13	A B C D E	23	A B C D E	33	A B C D E	43	A B C D E
4	A B C D E	14	A B C D E	24	A B C D E	34	A B C D E	44	A B C D E
5	A B C D E	15	A B C D E	25	A B C D E	35	A B C D E	45	A B C D E
6	A B C D E	16	A B C D E	26	A B C D E	36	A B C D E	46	A B C D E
7	A B C D E	17	A B C D E	27	A B C D E	37	A B C D E	47	A B C D E
8	A B C D E	18	A B C D E	28	A B C D E	38	A B C D E	48	A B C D E
9	A B C D E	19	A B C D E	29	A B C D E	39	A B C D E	49	A B C D E
10	A B C D E	20	A B C D E	30	A B C D E	40	A B C D E	50	A B C D E

Name ______________________ Date __________ Class __________

SAT Subject Test Practice Test I: Math Level IC

Time—60 minutes, 50 Questions

All questions in the Math Level 1 and Math Level 2 Tests are multiple-choice questions in which you are asked to choose the **BEST** response from the five choices offered. The directions for the tests are below:

Directions: For each of the following problems, decide which is the BEST of the choices given. If the exact numerical value is not one of the choices, select the choice that best approximates this value. Then fill in the corresponding oval on the answer sheet.

Notes:

1. A scientific or graphing calculator will be necessary for answering some (but not all) of the questions in this test. For each question you will have to decide whether or not you should use a calculator.
2. Level 1: The only angle measure used on this test is degree measure. Make sure your calculator is in the degree mode.
   Level 2: For some questions in this test you may have to decide whether your calculator should be in the radian mode or the degree mode.
3. Figures that accompany problems in this test are intended to provide information useful in solving the problems. They are drawn as accurately as possible EXCEPT when it is stated in a specific problem that its figure is not drawn to scale. All figures lie in a plane unless otherwise indicated.
4. Unless otherwise specified, the domain of any function f is assumed to be the set of all real numbers x for which $f(x)$ is a real number. The range of f is assumed to be the set of all real numbers $f(x)$, where x is in the domain of f.
5. Reference information that may be useful in answering the questions in this test can be found below.

Reference Information. The following information is for your reference in answering some of the questions in this test.

Volume of a right circular cone with radius r and height h: $V = \frac{1}{3}\pi r^2 h$

Lateral Area of a right circular cone with circumference of the base c and slant height ℓ: $S = \frac{1}{2}c\ell$

Volume of a sphere with radius r: $V = \frac{4}{3}\pi r^3$

Surface Area of a sphere with radius r: $S = 4\pi r^2$

Volume of a pyramid with base area B and height h: $V = \frac{1}{3}Bh$

USE THIS SPACE FOR SCRATCH WORK.

1. If $\frac{x-3}{4} = \frac{x-5}{7}$, what is the value of x?

(A) $\frac{1}{23}$

(B) $\frac{3}{23}$

(C) $\frac{1}{3}$

(D) 3

(E) $\frac{23}{3}$

GO ON

Name ________________________ Date __________ Class __________

SAT Subject Test Practice Test I: Math Level IC *continued*

USE THIS SPACE FOR SCRATCH WORK.

2. Triangle *ABC* is a right triangle with sides of length 4, 6, and *x*. If $4 < x < 6$, what is the approximate value of *x*?

(A) 4

(B) 4.47

(C) 5.21

(D) 5.63

(E) 7.21

3. What is the approximate length of the longer leg of triangle *EFG*?

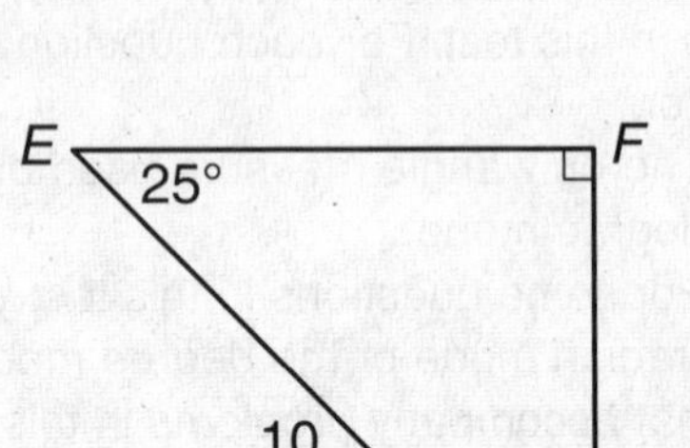

(A) 4.23

(B) 4.66

(C) 9.06

(D) 11.03

(E) 21.45

4. If $h(x) = x^2 - 3x + 1$, what is $[h(-2)]^2$?

(A) -9

(B) 5

(C) 11

(D) 81

(E) 121

5. John's six test scores are given. What is the positive difference between the mean of the scores and the median of the scores?

86, 72, 92, 62, 99, 93

(A) 4

(B) 5

(C) 7

(D) 85

(E) 89

GO ON

USE THIS SPACE FOR SCRATCH WORK.

6. What is the smallest integer that satisfies the inequality $4x + 5 > 2x - 3$?

(A) -5

(B) -4

(C) -3

(D) 4

(E) 5

7. If figure *ABCD* is a parallelogram, what is the *x*-coordinate of point *B*?

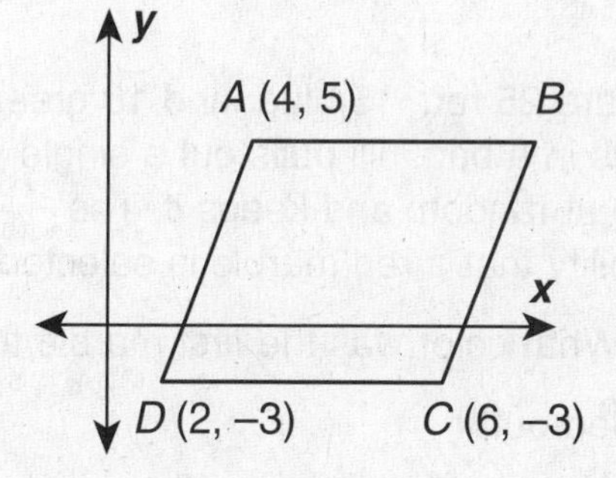

(A) 2

(B) 5

(C) 6

(D) 8

(E) 10

8. What is the fifth term of the arithmetic sequence 2, ___, 8, ___, ___, … ?

(A) 5

(B) 11

(C) 13

(D) 14

(E) 15

9. If the volume of a cone is 24π cubic feet and its height is 2 feet, what is the diameter of the base in feet?

(A) 3

(B) 6

(C) 12

(D) 24

(E) 36

GO ON

USE THIS SPACE FOR SCRATCH WORK.

10. If $3^{2x+1} = 81$, then $x =$

(A) 1

(B) $\frac{3}{2}$

(C) 2

(D) $\frac{5}{2}$

(E) 3

11. There are 25 red, 15 blue, and 10 green marbles in a bag. Jill pulls out a single marble at random and keeps it. The probability that a red marble is selected next is $\frac{24}{49}$. What color was the first marble that Jill pulled out?

(A) red

(B) blue

(C) green

(D) blue or green

(E) cannot be determined

12. If $f(x) = 2x - 6$, then $f^{-1}(x)$ is

(A) $6 - 2x$

(B) $\frac{1}{2}x - 6$

(C) $\frac{1}{2}x - 3$

(D) $\frac{1}{2}x + 3$

(E) $\frac{1}{2}x + 6$

13. What is the area of a parallelogram with vertices at (0, 0), (2, 3), (5, 0), and (7, 3)?

(A) 10

(B) 14

(C) 15

(D) 21

(E) 35

GO ON

Name ______________________ Date ____________ Class ____________

SAT Subject Test Practice Test I: Math Level IC *continued*

USE THIS SPACE FOR SCRATCH WORK.

14. $(3x + 4)^{-2} =$

(A) $9x^2 + 16$

(B) $9x^2 + \frac{1}{16}$

(C) $\frac{1}{9x^2 + 16}$

(D) $\frac{1}{9x^2 + 12x + 16}$

(E) $\frac{1}{9x^2 + 24x + 16}$

15. Simplify $\frac{\frac{x}{y}}{\frac{x}{y} + \frac{y}{x}}$.

(A) $\frac{1}{xy}$

(B) $\frac{1}{x^2}$

(C) $\frac{1}{y^2}$

(D) $\frac{x^2}{x^2 + y^2}$

(E) $\frac{xy}{xy + y^2}$

16. If the area of the triangle shown is 30 square units, what is the value of y?

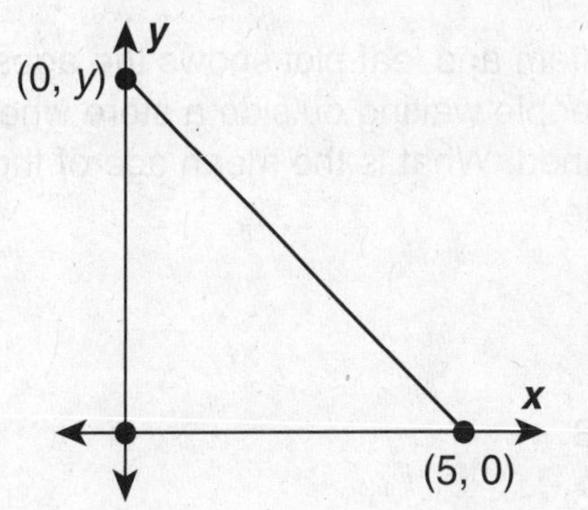

(A) 5

(B) 6

(C) 8

(D) 10

(E) 12

17. What is the surface area of a cylinder that has a top and bottom if the height is 8.5 units and the diameter is 3 units?

(A) 18π square units

(B) 30π square units

(C) 32π square units

(D) 36π square units

(E) 54π square units

GO ON

SAT Subject Test Practice Test I: Math Level IC *continued*

USE THIS SPACE FOR SCRATCH WORK.

18. What is the approximate length of segment AC?

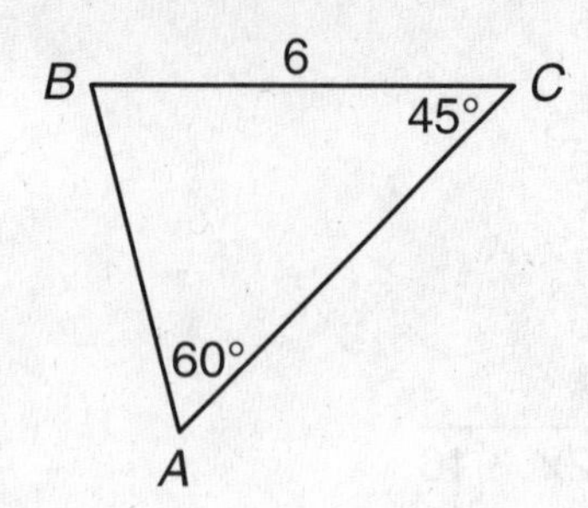

(A) 4.9

(B) 6.7

(C) 6.9

(D) 7.1

(E) 7.3

19. If $x = -3$ and $x^2 + xy = 15$, what is the value of y?

(A) -2

(B) -1

(C) 2

(D) 3

(E) 5

20. If $x = a + b$ then $2^x + 2^a \cdot 2^b =$

(A) $2 + 2^{a+b}$

(B) $2^{a+b} + 2^{ab}$

(C) 2^{2a+2b}

(D) 2^{a+b+1}

(E) 2^{a+b+ab}

21. The stem and leaf plot shows the ages of the people waiting outside a store when it opened. What is the mean age of the people?

Stem	Leaf
1	1 2 3
2	1 3 5
3	0 1 2

(A) 2

(B) 3

(C) 18

(D) 22

(E) 23

GO ON

SAT Subject Test Practice Test I: Math Level IC *continued*

USE THIS SPACE FOR SCRATCH WORK.

22. What is the ratio of the area of the square to the area of the circle if the length of a side of the square is 10 units?

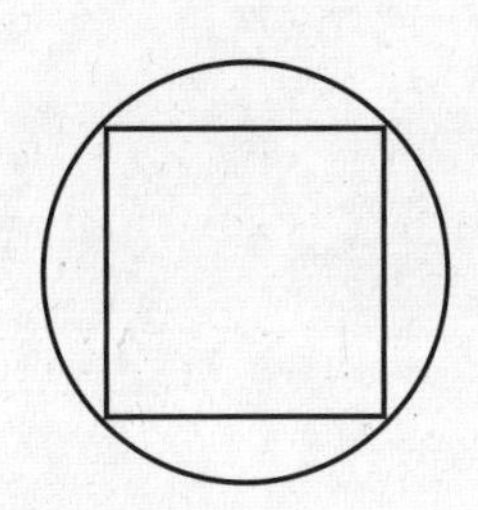

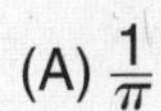

(A) $\frac{1}{\pi}$

(B) $\frac{\sqrt{3}}{\pi}$

(C) $\frac{2}{\pi}$

(D) $\frac{4}{\pi}$

(E) $\frac{5}{\pi}$

23. If $f(x) = x^2 - x$ and $g(x) = 3x + 2$, what is $g(f(1)) - 2$?

(A) 0

(B) 1

(C) 2

(D) 3

(E) 18

24. Which of the following is equivalent to $\log\left(\frac{1}{10^a}\right)$

(A) -1

(B) $-a$

(C) a

(D) 10^{-a}

(E) 10^a

25. What is the area of the circle whose equation is $(x - 3)^2 + (y + 5)^2 = 18$?

(A) 9π

(B) 18π

(C) 72π

(D) 81π

(E) 324π

GO ON

USE THIS SPACE FOR SCRATCH WORK.

26. What is the tenth term of the arithmetic sequence whose first term is x and whose third term is $x + 6a$?

(A) $33a$

(B) $x + 24a$

(C) $x + 27a$

(D) $x + 30a$

(E) $x + 33a$

27. Figure $ABCD$ is a trapezoid with area 50 square units. If the length of $\overline{AB}$ is 6 units and the length of $\overline{BC}$ is 5 units, what is the length of $\overline{CD}$?

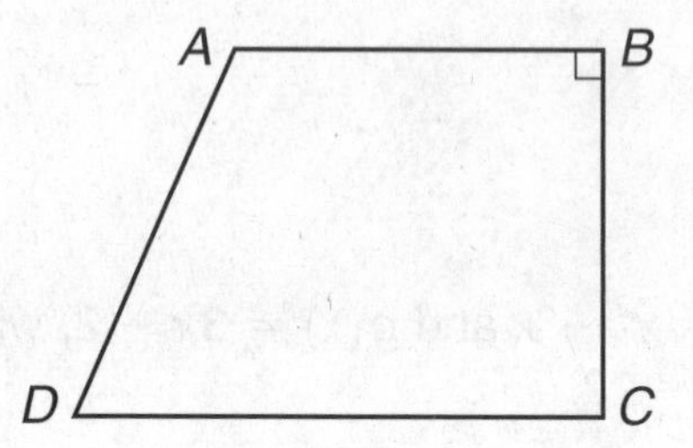

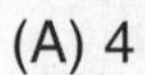

(A) 4

(B) 5

(C) 6

(D) 10

(E) 14

28. A root of $4x^3 + 12x^2 + 9x + 27 = 0$ is $(2x - 3i)$. Which of the following is <u>another</u> root?

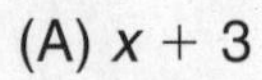

(A) $x + 3$

(B) $x - 3$

(C) $2x - 2i$

(D) $2x + 2i$

(E) $2x - 3$

29. The points (0, 0), (16, 0), (12, 12), and (4, 12) are the corners of a

(A) rhombus

(B) rectangle

(C) parallelogram

(D) trapezoid

(E) None of the above

GO ON

Name ______________________ Date ____________ Class ____________

SAT Subject Test Practice Test I: Math Level IC *continued*

USE THIS SPACE FOR SCRATCH WORK.

30. A sphere is inscribed inside a cube. What is that probability that a point that is inside the cube is also inside the sphere?

(A) $\frac{1}{\pi}$

(B) $\frac{\pi}{8}$

(C) $\frac{\pi}{6}$

(D) $\frac{\pi}{4}$

(E) $\frac{2}{\pi}$

31. Solve for x: $\sqrt{ax + 1} - \sqrt{ax - 1} = \sqrt{ax}$.

(A) $-\frac{2}{3a}$

(B) $\frac{2}{3a}$

(C) $\frac{\pm 2\sqrt{3}}{3a}$

(D) $\frac{2\sqrt{3}}{3a}$

(E) $-\frac{2\sqrt{3}}{3a}$

32. Find the approximate value of m so that the area of triangle ABC is equal to the area of trapezoid $BCDO$.

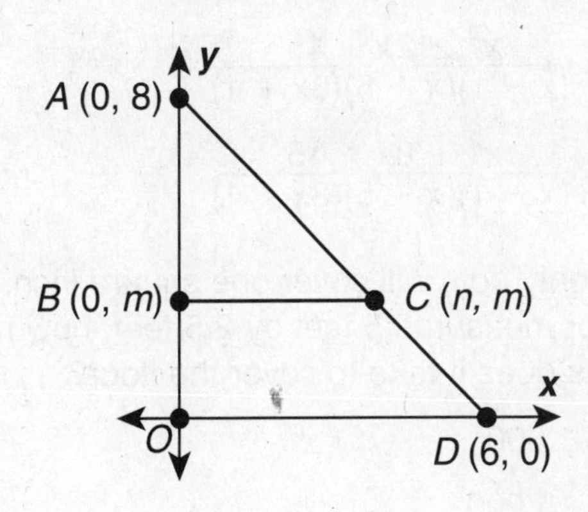

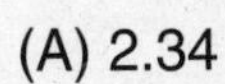

(A) 2.34

(B) 2.88

(C) 3.14

(D) 3.34

(E) 3.88

33. If $f(x) = 3x + 5$ and $f(g(x)) = 6x - 4$, what is $g(x)$?

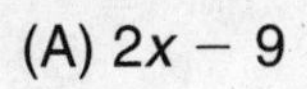

(A) $2x - 9$

(B) $2x - 3$

(C) $3x - 9$

(D) $3x - 3$

(E) $9x + 1$

GO ON

SAT Subject Test Practice Test I: Math Level IC *continued*

USE THIS SPACE FOR SCRATCH WORK.

34. Points A and C lie on a straight road and point B lies directly above the road. The angle of elevation from point A to point B is 35° and the angle of depression from point B to point C is 35°. If the distance from A to C is 20 miles, approximately how many miles above the road is point B?

(A) 6.25

(B) 6.75

(C) 7.00

(D) 7.50

(E) 8.50

35. If $\dfrac{2x-3}{3x^2+16x+5} + A = \dfrac{3x^2+3x+18}{3x^3+13x^2-11x-5}$, then $A =$

(A) $\dfrac{x-5}{(x+5)(x-1)}$

(B) $\dfrac{x+3}{(x-1)(3x+1)}$

(C) $\dfrac{x-3}{(x+5)(3x+1)}$

(D) $\dfrac{x^2-2x+15}{(x-1)(x+5)(3x+1)}$

(E) $\dfrac{x^2+8x-15}{(x-1)(x+5)(3x+1)}$

36. Twenty dots will cover one square inch. If a floor measures 5 feet by 2.5 feet, how many dots does it take to cover the floor?

(A) 1,800

(B) 15,000

(C) 36,000

(D) 45,000

(E) 60,000

GO ON

Name ______________________ Date __________ Class __________

SAT Subject Test Practice Test I: Math Level IC *continued*

USE THIS SPACE FOR SCRATCH WORK.

37. The zeros of $m(x) = \dfrac{x^2 + 3x + 2}{x^2 - 3x + 2}$ are the first two terms of a sequence. Each term in the sequence is found by adding the two terms before it. If each term is smaller than the one before it, what is the fifth term of the sequence?

(A) −8

(B) −5

(C) 7

(D) 8

(E) 9

38. If *p*, *q*, and *r* are all different integer factors of 48, then the greatest value of the product of *p*, *q*, and *r* is

(A) 2,304

(B) 4,608

(C) 13,824

(D) 18,432

(E) 110,592

39. The figure shows a semicircle on top of an isosceles right triangle. If the length of $\overset{\frown}{AB}$ is 16π, what is the approximate length of $\overline{BC}$?

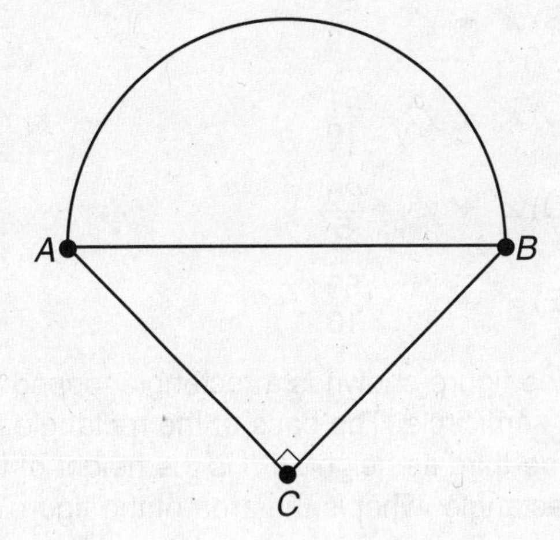

(A) 2.8

(B) 4.0

(C) 5.7

(D) 11.3

(E) 22.6

40. If $g(x) = \dfrac{5x - 3}{2x^2 - 11 - 6}$, what is the sum of all the real numbers that are not in the domain of $g(x)$?

(A) −2

(B) 0.5

(C) 2

(D) 5.5

(E) 6.5

GO ON

USE THIS SPACE FOR SCRATCH WORK.

41. If $i^2 = -1$, then $i^{162} =$

(A) $-i$

(B) -1

(C) 0

(D) 1

(E) i

42. When $(2x - 3)^5$ is written in the form $a_1x^5 + a_2x^4 + a_3x^3 + \ldots$, the sum of the first three coefficients is

(A) 16

(B) 56

(C) 152

(D) 512

(E) 992

43. What is an equation of the circle that has its center at the origin and is tangent to the line $y = -3x + 7$?

(A) $x^2 + y^2 = 5$

(B) $x^2 + y^2 = \frac{49}{10}$

(C) $x^2 + y^2 = \frac{51}{10}$

(D) $x^2 + y^2 = \frac{26}{5}$

(E) $x^2 + y^2 = \frac{53}{10}$

44. The figure shown is a rectangle topped with a semicircle. The base of the rectangle is one-third its height. If h is the height of the rectangle, what is the area of the figure?

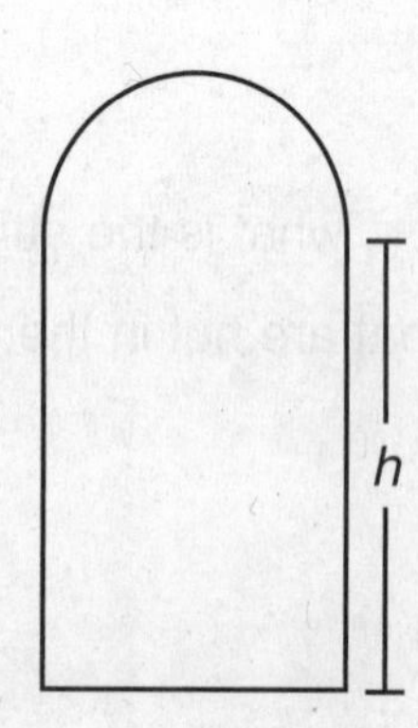

(A) $\left(\frac{24 + \pi}{72}\right)h^2$

(B) $\left(\frac{12 + \pi}{36}\right)h^2$

(C) $\left(\frac{\pi}{3}\right)h^2$

(D) $\left(\frac{24 + 9\pi}{8}\right)h^2$

(E) $\left(\frac{12 + 9\pi}{4}\right)h^2$

SAT Subject Test Practice Test I: Math Level IC *continued*

USE THIS SPACE FOR SCRATCH WORK.

45. Which of the following could be the factors of $x - h$?

(A) $(x^{\frac{1}{2}} - h^{\frac{1}{2}})^2$

(B) $(x^{\frac{1}{2}} + h^{\frac{1}{2}})^2$

(C) $(x^{\frac{1}{3}} - h^{\frac{1}{3}})(x^{\frac{1}{3}} + x^{\frac{1}{3}}h^{\frac{1}{3}} + h^{\frac{1}{3}})$

(D) $(x^{\frac{1}{3}} - h^{\frac{1}{3}})(x^{\frac{2}{3}} + x^{\frac{1}{3}}h^{\frac{1}{3}} + h^{\frac{2}{3}})$

(E) $(x^{\frac{1}{3}} - h^{\frac{1}{3}})(x^{\frac{2}{3}} + x^{\frac{2}{3}}h^{\frac{2}{3}} + h^{\frac{2}{3}})$

46. If $h(x) = (f \circ g)(x)$ and $h^{-1}(x)$ is the inverse of $h(x)$, then which of the following must be equal to x?

(A) $(h^{-1} \circ f \circ g)(x)$

(B) $(h^{-1} \circ g \circ f)(x)$

(C) $(f^{-1} \circ g^{-1} \circ h^{-1})(x)$

(D) $(g^{-1} \circ f^{-1} \circ h^{-1})(x)$

(E) Not enough information

47. The measure of $\overparen{ABC}$ is 280° and the length of $\overline{AD}$ is 10 units. What is the approximate length of $\overline{AC}$?

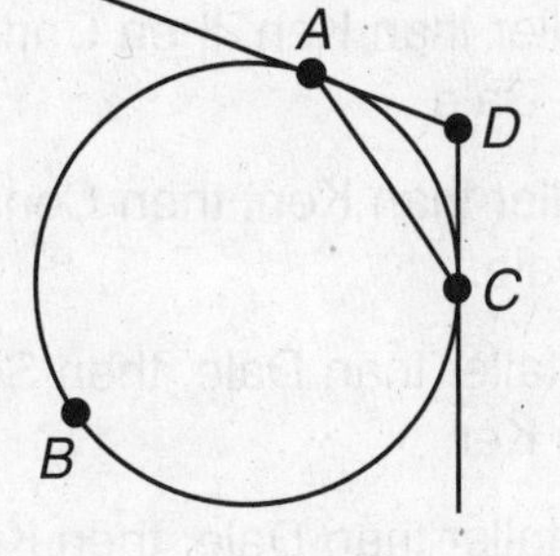

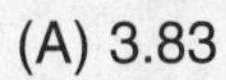

(A) 3.83

(B) 7.66

(C) 10.44

(D) 12.91

(E) 15.32

48. If $y = rx^2 + sx + t$, where r, s, and t are real numbers such that $|s| < 1$, $|r| > |t| > 1$, and $r \cdot t > 0$, what is true about the zeros of y?

(A) There is one real zero.

(B) There are two real zeros.

(C) There is one complex zero.

(D) There are two complex zeros.

(E) Not enough information

GO ON

USE THIS SPACE FOR SCRATCH WORK.

49. Figure *ABCD* is a rectangle whose length is twice its width. $\overset{\frown}{FC}$ and $\overset{\frown}{AE}$ are arcs of circles centered at *B* and *D* respectively. If the length of $\overline{AD}$ is *x*, then the area of the shade region is

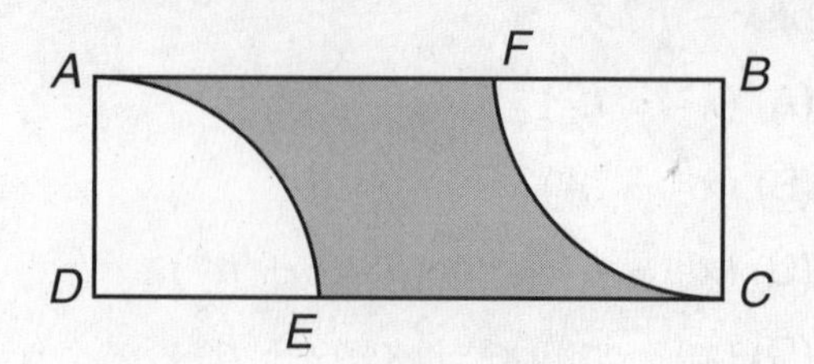

(A) $\left(\frac{2-\pi}{4}\right)x^2$

(B) $\left(\frac{4-\pi}{2}\right)x^2$

(C) $\left(\frac{\pi-2}{4}\right)x^2$

(D) $\left(\frac{2}{\pi}\right)x^2$

(E) $\left(\frac{4}{\pi}\right)x^2$

50. If Ken is taller than Scott, then Dale is shorter than Connie. Which of the following must be true?

(A) If Scott is taller than Ken, then Connie is shorter than Dale.

(B) If Scott is taller than Ken, then Connie is taller than Dale.

(C) If Connie is taller than Dale, then Scott is taller than Ken.

(D) If Connie is taller than Dale, then Ken is taller than Scott.

(E) If Dale is taller than Connie, then Scott is taller than Ken.

STOP **If you finish before time is called, you may check your work on this section only. Do not turn to any other section in the test.**

Name ______________________________ Date ___________ Class ___________

SAT Subject Test Practice Test II: Math Level IC

Time—60 minutes, 50 Questions

All questions in the Math Level 1 and Math Level 2 Tests are multiple-choice questions in which you are asked to choose the **BEST** response from the five choices offered. The directions for the tests are below:

Directions: For each of the following problems, decide which is the BEST of the choices given. If the exact numerical value is not one of the choices, select the choice that best approximates this value. Then fill in the corresponding oval on the answer sheet.

Notes:

1. A scientific or graphing calculator will be necessary for answering some (but not all) of the questions in this test. For each question you will have to decide whether or not you should use a calculator.
2. Level 1: The only angle measure used on this test is degree measure. Make sure your calculator is in the degree mode.
   Level 2: For some questions in this test you may have to decide whether your calculator should be in the radian mode or the degree mode.
3. Figures that accompany problems in this test are intended to provide information useful in solving the problems. They are drawn as accurately as possible EXCEPT when it is stated in a specific problem that its figure is not drawn to scale. All figures lie in a plane unless otherwise indicated.
4. Unless otherwise specified, the domain of any function f is assumed to be the set of all real numbers x for which $f(x)$ is a real number. The range of f is assumed to be the set of all real numbers $f(x)$, where x is in the domain of f.
5. Reference information that may be useful in answering the questions in this test can be found below.

Reference Information. The following information is for your reference in answering some of the questions in this test.

Volume of a right circular cone with radius r and height h: $V = \frac{1}{3}\pi r^2 h$

Lateral Area of a right circular cone with circumference of the base c and slant height ℓ: $S = \frac{1}{2}c\ell$

Volume of a sphere with radius r: $V = \frac{4}{3}\pi r^3$

Surface Area of a sphere with radius r: $S = 4\pi r^2$

Volume of a pyramid with base area B and height h: $V = \frac{1}{3}Bh$

USE THIS SPACE FOR SCRATCH WORK.

1. If $3x - 2 = y$, which of the following is equal to $y - 2$?

(A) $3x$

(B) $3x - 4$

(C) $3x + 2$

(D) $3x + 4$

(E) $3x + 8$

GO ON

Name ______________________ Date ____________ Class ____________

SAT Subject Test Practice Test II: Math Level IC *continued*

USE THIS SPACE FOR SCRATCH WORK.

2. What is the circumference of a circle with center (4, 0) if the circle passes through the point (4, −3)?

(A) π

(B) $\pi\sqrt{3}$

(C) 3π

(D) 6π

(E) 9π

3. Three sets of data are given below. What is the median of the 3 ranges?

Data *A*: −3, 2, −5, 6, 8
Data *B*: 4, 9, 6, 2
Data *C*: 4, 5, 6, 7, 12, 9

(A) 2

(B) 5

(C) 7

(D) 8

(E) 13

4. If $\frac{5x + 8}{9} = \frac{a}{9} + \frac{2}{9}$, which of the following is equal to a?

(A) $5x$

(B) $5x - 6$

(C) $5x - 2$

(D) $5x + 2$

(E) $5x + 6$

5. What is the distance between points (2, 3) and (−5, 7) rounded to two decimal places?

(A) 5.00

(B) 8.06

(C) 10.44

(D) 12.04

(E) 12.21

GO ON

USE THIS SPACE FOR SCRATCH WORK.

6. If $f(x) = x^2 + 2x + 3$ then $-f(-3) =$

(A) -18

(B) -12

(C) -6

(D) 6

(E) 18

7. At a certain school, the probability of being male is 60% and the probability of having red hair is 10%. What is the probability of being a female with red hair?

(A) 4%

(B) 6%

(C) 10%

(D) 36%

(E) 40%

8. If $5^{7x-5} = 25$, then $x =$

(A) -1

(B) $-\frac{3}{7}$

(C) $\frac{3}{7}$

(D) 1

(E) 2

9. If the lengths of the sides of a square are integers, which of the following could be the area of the square in square units?

(A) 1

(B) 2

(C) 6

(D) 8

(E) All of the above

GO ON

SAT Subject Test Practice Test II: Math Level IC *continued*

USE THIS SPACE FOR SCRATCH WORK.

10. The length of the base of a certain television screen is 50 inches and it makes a $35°$ angle with the diagonal of the screen. Approximately how long, in inches, is the diagonal?

(A) 61

(B) 62

(C) 63

(D) 64

(E) 65

11. If $(2, -1)$ is the solution to the system of equations shown below, what is the value of b?

$$2x + y = a$$

$$3x - y = b$$

(A) -5

(B) 1

(C) 3

(D) 5

(E) 7

12. For $a = 0, 1, 2, 3,$ and 4 only, $h(a) = a^2 + 1$. What is the sum of all the values of $h(a)$?

(A) 1

(B) 10

(C) 17

(D) 35

(E) Cannot be found

13. What is the equation of the figure shown?

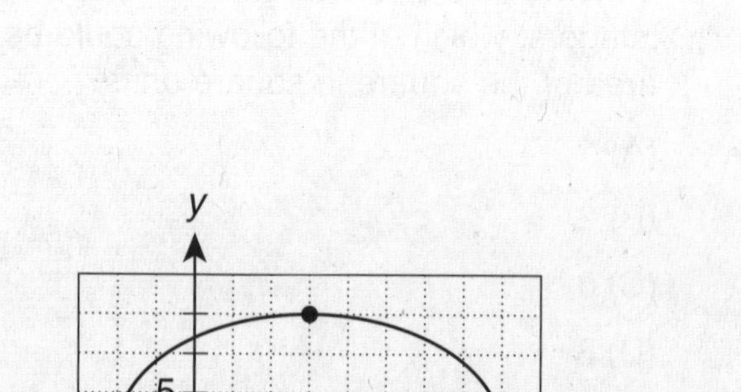

(A) $\frac{x^2}{25} + \frac{y^2}{9} = 1$

(B) $\frac{x^2}{64} + \frac{y^2}{16} = 1$

(C) $\frac{(x-3)^2}{25} + \frac{(y-4)^2}{9} = 1$

(D) $\frac{(x+3)^2}{25} + \frac{(y+4)^2}{9} = 1$

(E) $\frac{(x+3)^2}{64} + \frac{(y+4)^2}{16} = 1$

GO ON

USE THIS SPACE FOR SCRATCH WORK.

14. If $x = 6$, what is the value of $(x^{\frac{1}{6}})(x^{\frac{1}{3}})(x^{\frac{1}{2}})$?

(A) $\frac{1}{36}$

(B) $\frac{1}{6}$

(C) 1

(D) 6

(E) 36

15. If $f(x) = 3x - 2$ and $g(x) = x^2$, find $f(g(-3))$.

(A) -11

(B) 3

(C) 9

(D) 25

(E) 121

16. The lengths of two sides of a triangle are 9 units and 14 units. Which of the following could NOT be the length of the third side?

(A) 5

(B) 7

(C) 10

(D) 17

(E) 22

17. If $f(x) = 3x + 2$ and $g(x) = 5x - 8$, for what values of x is the difference between $f(x)$ and $g(x)$ greater than 0?

(A) $x > 5$

(B) $x < 5$

(C) $x > \frac{3}{4}$

(D) $x < -3$

(E) $x > -3$

GO ON

USE THIS SPACE FOR SCRATCH WORK.

18. Circle C has its center at (2, 3). What is the slope of the tangent line to circle C at the point (5, −1)?

(A) $-\frac{4}{3}$

(B) $-\frac{1}{5}$

(C) $\frac{1}{5}$

(D) $\frac{3}{4}$

(E) 5

19. For what values of x is $|2x + 1| + 3 < 8$?

(A) $x < 2$

(B) $-2 < x < 3$

(C) $-3 < x < 2$

(D) $x < -2$ or $x > 3$

(E) $x < -3$ or $x > 2$

20. The lengths of the sides of a rectangle are consecutive even integers. If the perimeter of the rectangle is 28 units, what is the area of the rectangle in square units?

(A) 14

(B) 24

(C) 28

(D) 32

(E) 48

21. For which of the x-values given is the graph of $y = -x^3 - x^2$ above the x-axis?

(A) −2

(B) −1

(C) 0

(D) 1

(E) 2

GO ON

Name ______________________ Date __________ Class __________

SAT Subject Test Practice Test II: Math Level IC *continued*

USE THIS SPACE FOR SCRATCH WORK.

22. If $4m^2 + 9n^2 = 1$ and $(2m - 3n)^2 = 13$, what is the value of mn?

(A) -12

(B) -2

(C) -1

(D) 1

(E) 12

23. In the diagram, the measure of angle A is

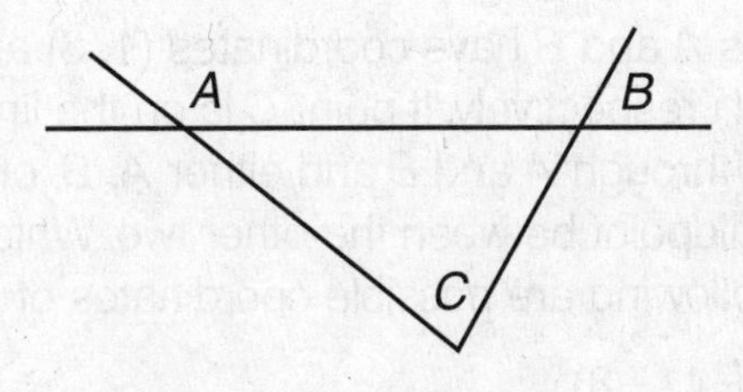

(A) $B - C$

(B) $B + C$

(C) $180 - (B + C)$

(D) $180 - (B - C)$

(E) $180 - (C - B)$

24. Which of the following is the domain of the function $g(x) = \dfrac{\sqrt{x - 5}}{6 - x}$?

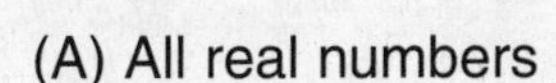

(A) All real numbers

(B) $5 \le x < 6 \cup x > 6$

(C) $x \ge 5$

(D) $x \ge 6$

(E) $x \ge 7$

25. The sequence S_n has the following properties: $a_1 = -3$, $a_2 = -2$, and $a_n = (a_{n-1})(a_{n-2})$. What is the sum of the first four terms of the sequence?

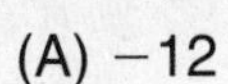

(A) -12

(B) -11

(C) 6

(D) 786

(E) 864

26. The approximate value of the smallest angle that the line shown makes with the x-axis is

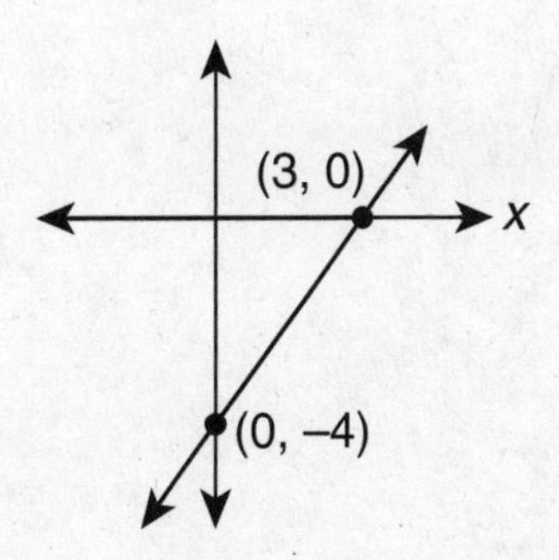

(A) 37°

(B) 43°

(C) 47°

(D) 53°

(E) 57°

GO ON

USE THIS SPACE FOR SCRATCH WORK.

27. If $2x^2 + 6x + c = 0$ has exactly one solution, what is the value of c?

(A) -9

(B) -4.5

(C) 0

(D) 4.5

(E) 9

28. Points A and B have coordinates (1, 3) and (5, 11) respectively. If point C is on the line that goes through A and B and either A, B, or C is the midpoint between the other two. Which of the following are possible coordinates of C?

(A) $(-4, -8)$

(B) (4, 8)

(C) (6, 4)

(D) (6, 14)

(E) (9, 19)

29. If $h(x) = \dfrac{x^2 - 9}{x^2 + x - 12}$, what are the zeros of $h(x)$?

(A) $x = -4, x = -3, x = 3$

(B) $x = -4, x = 3$

(C) $x = -3, x = 3$

(D) $x = -3$

(E) $x = 3$

30. If the area of $\odot P = A$ and the area of $\odot M = B$, then $A =$

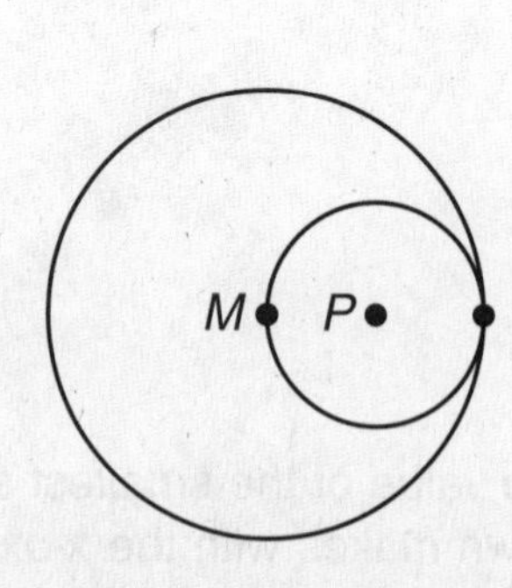

(A) $\frac{1}{8}B$

(B) $\frac{1}{4}B$

(C) $\frac{1}{3}B$

(D) $\frac{1}{2}B$

(E) B

GO ON

SAT Subject Test Practice Test II: Math Level IC *continued*

USE THIS SPACE FOR SCRATCH WORK.

31. If $x = -2$ is a solution to the equation $x^2 + bx + c = 0$ and $c = 3b$, what is the value of c?

(A) -12

(B) -9

(C) -6

(D) 6

(E) 12

32. Which of the following points lie inside the circle with center $(2, -3)$ and radius 5?

(A) $(-2, 0)$

(B) $(-2, 1)$

(C) $(3, -7)$

(D) $(3, 3)$

(E) $(5, 1)$

33. Ms. Gomez said there are 10 boys and 15 girls in her class. If Ms. Gomez is incorrect, then which of the following MUST be correct?

(A) There are not 10 boys in her class.

(B) There are either 10 boys or 15 girls in her class.

(C) There are not 25 people in her class.

(D) There are not 10 boys in her class and there are not 15 girls in her class.

(E) If there are 10 boys in her class, then there are not 15 girls in her class.

34. A cylinder and a cone both have the same height and volume. If the radius of the base of the cone is 12 units, what is the radius of the cylinder?

(A) $2\sqrt{3}$

(B) $3\sqrt{2}$

(C) $3\sqrt{3}$

(D) $4\sqrt{2}$

(E) $4\sqrt{3}$

GO ON

USE THIS SPACE FOR SCRATCH WORK.

35. In the triangle shown, $DE \parallel BC$, the length of DE is 6 units, the length of BC is 18 units, and the area of $\triangle ABC$ is 657 square units. What is the area of $\triangle ADE$ in square units?

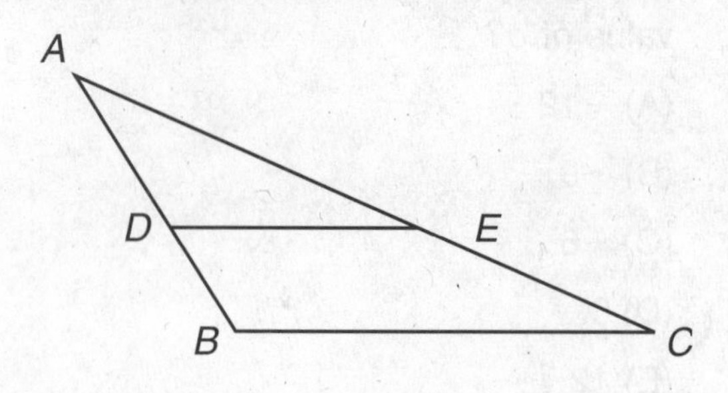

(A) 67

(B) 73

(C) 109

(D) 127

(E) 219

36. If $5^{2a} = 7^{b}$, what is the approximate ratio of a to b?

(A) 0.06

(B) 0.3

(C) 0.6

(D) 1.2

(E) 1.5

37. Which of the following has its highest point at $(-3, -4)$?

(A) $y = -(x + 3) - 4$

(B) $y = -x^2 - 6x - 13$

(C) $y = -x^2 - 6x - 5$

(D) $y = x^2 + 6x + 5$

(E) $y = x^2 + 6x + 13$

38. Find a value of k so that $(x + k)$ is a factor of $3x^2 + 11x + k + 8$.

(A) -2

(B) -1

(C) 0

(D) 1

(E) 2

GO ON

USE THIS SPACE FOR SCRATCH WORK.

39. If $y = \frac{a(x-b)^2 + c}{d}$, then $x =$

(A) $b + \frac{\sqrt{d(y-c)}}{a}$

(B) $b \pm \frac{\sqrt{d(y-c)}}{a}$

(C) $b - \sqrt{\frac{dy-c}{a}}$

(D) $b \pm \sqrt{\frac{dy-c}{a}}$

(E) $b + \sqrt{\frac{dy-c}{a}}$

40. The figure shown is a rhombus with sides of length 2 units. If the longer diagonal makes a 30° angle with one of the sides, then how long is the shorter diagonal?

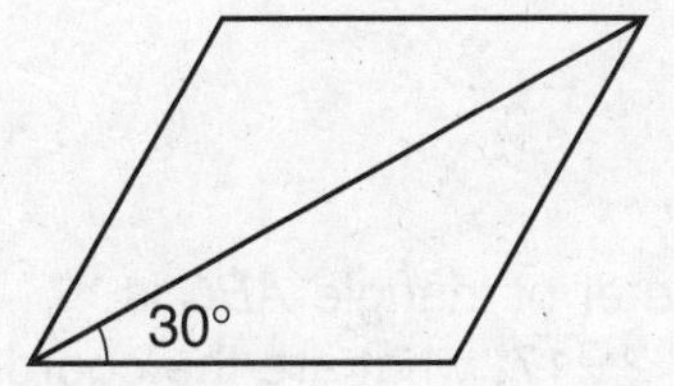

(A) 1

(B) $\sqrt{3}$

(C) 2

(D) 3

(E) $2\sqrt{3}$

41. A cylinder with a top and bottom is inscribed in a cube that has length 6 units. What is the approximate surface area of the cylinder in square units?

(A) 84.82

(B) 113.10

(C) 169.65

(D) 226.19

(E) 452.92

42. If the square of the sum of x and y is 50 and the square of the difference of x and y is 30, what is the product of x and y?

(A) 5

(B) 10

(C) 20

(D) 30

(E) 600

GO ON

Name ______________________ Date __________ Class __________

USE THIS SPACE FOR SCRATCH WORK.

43. Mr. Smith has three boxes. Each box contains 26 cards, each with a different letter of the alphabet written on it. If you choose one card from each box, what is the probability that you choose all vowels?

(A) $\frac{1}{26^3}$

(B) $\frac{5^3}{26^3}$

(C) $\frac{1}{26}$

(D) $\frac{3}{26}$

(E) $\frac{5}{26}$

44. If the perimeter of triangle ABC is $6 + 4\sqrt{5} + 2\sqrt{17}$, what are the coordinates of point B?

B

A (−2, −3)

C (4, −3)

(A) (0, 2)

(B) (0, 3)

(C) (0, 4)

(D) (0, 5)

(E) (0, 6)

45. What is the 152nd digit to the right of the decimal point if $\frac{152}{333}$ is written as a decimal?

(A) 3

(B) 4

(C) 5

(D) 6

(E) 7

GO ON

SAT Subject Test Practice Test II: Math Level IC *continued*

USE THIS SPACE FOR SCRATCH WORK.

46. The lengths of the legs of triangle *PRS* are 5 units and 12 units long. If $\overline{PS}$ is the diameter of the circle, what is the area of square *ABCD* in square units?

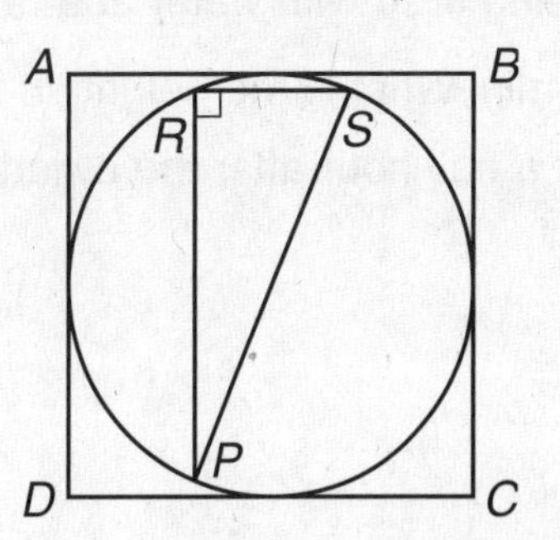

(A) 30

(B) 60

(C) 144

(D) 169

(E) 300

47. The sum of the squares of four consecutive odd integers is 36. What is the smallest possible value for the least of the four integers?

(A) −7

(B) −5

(C) −3

(D) −1

(E) 1

48. The radii of circles *O* and *P* are 5 units and 3 units respectively. $\overline{AC}$ is tangent to circles *O* and *P* at points *A* and *B* respectively. What is the ratio of *AB to BC*?

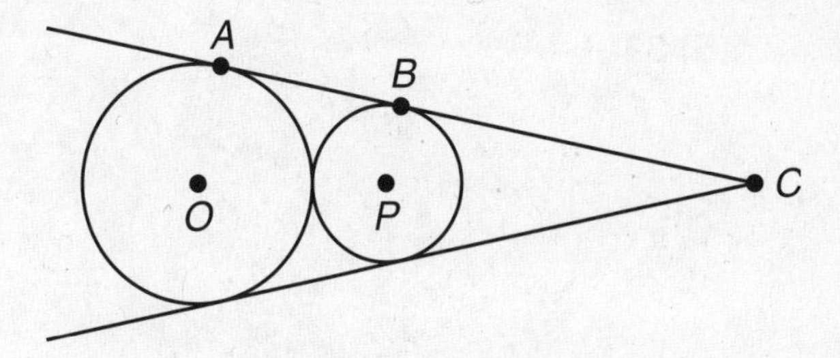

(A) $\frac{2}{5}$

(B) $\frac{3}{5}$

(C) $\frac{2}{3}$

(D) $\frac{5}{3}$

(E) $\frac{5}{2}$

GO ON

Name ______________________ Date __________ Class __________

SAT Subject Test Practice Test II: Math Level IC *continued*

USE THIS SPACE FOR SCRATCH WORK.

49. A cylinder with radius 5 feet and height 12 feet is being filled with water at a rate of $\frac{20\pi}{3}$ ft^3 per min. What is the height, in feet, of the water a half hour after the cylinder starts filling?

(A) 6

(B) 8

(C) 9

(D) 10

(E) 12

50. $\overline{AB}$ is tangent to $\odot O$ at point A, the length of $\overline{AB}$ is 50 units, and the length of $\overline{BC}$ is 30 units. What is the approximate measure, in degrees, of $\angle ABC$?

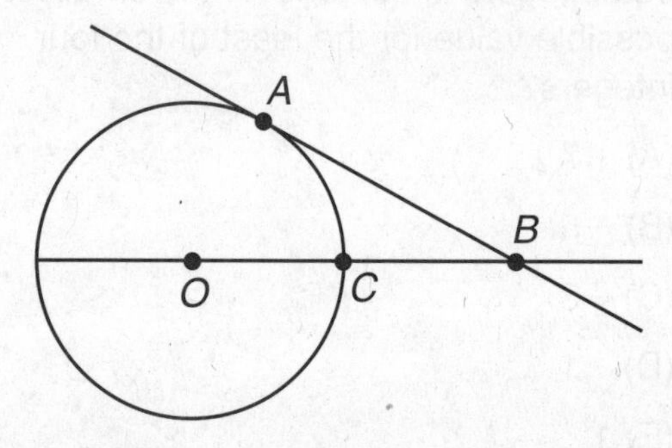

(A) 28

(B) 29

(C) 30

(D) 31

(E) 32

STOP **If you finish before time is called, you may check your work on this section only. Do not turn to any other section in the test.**

Name ______________________________ Date ___________ Class ___________

SAT Subject Test Practice Test I: Math Level IIC

Time—60 minutes, 50 Questions

All questions in the Math Level 1 and Math Level 2 Tests are multiple-choice questions in which you are asked to choose the **BEST** response from the five choices offered. The directions for the tests are below:

Directions: For each of the following problems, decide which is the BEST of the choices given. If the exact numerical value is not one of the choices, select the choice that best approximates this value. Then fill in the corresponding oval on the answer sheet.

Notes:

1. A scientific or graphing calculator will be necessary for answering some (but not all) of the questions in this test. For each question you will have to decide whether or not you should use a calculator.
2. Level 1: The only angle measure used on this test is degree measure. Make sure your calculator is in the degree mode.
   Level 2: For some questions in this test you may have to decide whether your calculator should be in the radian mode or the degree mode.
3. Figures that accompany problems in this test are intended to provide information useful in solving the problems. They are drawn as accurately as possible EXCEPT when it is stated in a specific problem that its figure is not drawn to scale. All figures lie in a plane unless otherwise indicated.
4. Unless otherwise specified, the domain of any function f is assumed to be the set of all real numbers x for which $f(x)$ is a real number. The range of f is assumed to be the set of all real numbers $f(x)$, where x is in the domain of f.
5. Reference information that may be useful in answering the questions in this test can be found on the page preceding Question 1.

Reference Information. The following information is for your reference in answering some of the questions in this test.

Volume of a right circular cone with radius r and height h: $V = \frac{1}{3}\pi r^2 h$

Lateral Area of a right circular cone with circumference of the base c and slant height ℓ: $S = \frac{1}{2}c\ell$

Volume of a sphere with radius r: $V = \frac{4}{3}\pi r^3$

Surface Area of a sphere with radius r: $S = 4\pi r^2$

Volume of a pyramid with base area B and height h: $V = \frac{1}{3}Bh$

USE THIS SPACE FOR SCRATCH WORK.

1. Which of the following is NOT a possible rational zero of $P(x) = 3x^3 + 2x^2 + 4x - 6$?

(A) $\frac{1}{6}$

(B) $\frac{1}{3}$

(C) $\frac{2}{3}$

(D) 1

(E) 6

GO ON

Name ______________________ Date __________ Class __________

SAT Subject Test Practice Test I: Math Level IIC *continued*

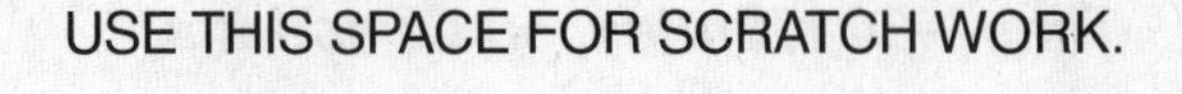

2. If $f(x)$ has a y-intercept at the point (0, 3), which of the following points must lie on the graph of $f(x + 4) - 1$?

(A) $(-4, 4)$

(B) $(-4, 2)$

(C) $(0, 3)$

(D) $(4, 2)$

(E) $(4, 4)$

3. $\sum_{k=2}^{5} (2k + 1) = ?$

(A) 1

(B) 2

(C) 5

(D) 11

(E) 32

4. The diagram represents three ships at sea, A, B, and C. Ship A is 25 miles from ship B and 35 miles from ship C and the measure of the angle from A to B to C is 20°. What is the measure of the angle from A to C to B, to the nearest degree?

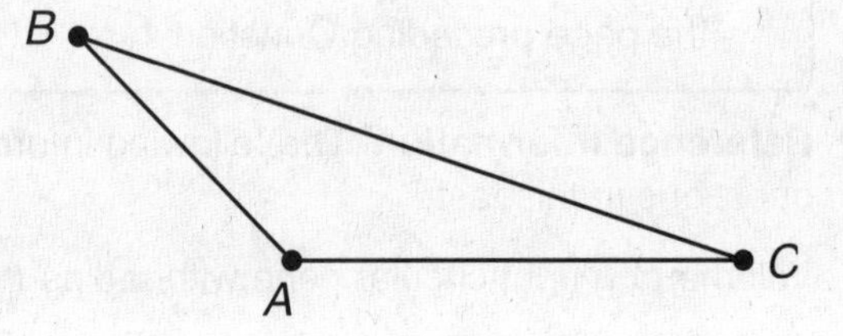

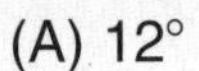

(A) 12°

(B) 14°

(C) 36°

(D) 44°

(E) 46°

5. $7i^7 + 8i^8 + 9i^9 + 10i^{10} = ?$

(A) $-2 - 2i$

(B) $-2i$

(C) $2i$

(D) $-2 + 2i$

(E) $2 + 2i$

GO ON

Name ______________________ Date __________ Class __________

SAT Subject Test Practice Test I: Math Level IIC *continued*

USE THIS SPACE FOR SCRATCH WORK.

6. If $f(x, y) = \frac{xy}{3}$ and $g(x) = 3^x$, the value of $g(f(1, 6))$ is

(A) 1

(B) 2

(C) 3

(D) 6

(E) 9

7. Find $f(3)$ if $f(x) = \begin{cases} 2 - x, & \text{if } x > 3 \\ x + 2, & \text{if } x \leq 3 \end{cases}$.

(A) –1

(B) –1 and 5

(C) 3

(D) 5

(E) $f(3)$ is undefined

8. What is the ratio of the surface area to the volume of a sphere whose diameter is 6 meters?

(A) $\frac{1}{1 \text{ meter}}$

(B) $\frac{1 \text{ meter}}{1}$

(C) $\frac{1}{1}$

(D) $\frac{1}{3 \text{ meter}}$

(E) $\frac{1 \text{ meter}}{3}$

9. What is the z-coordinate of the midpoint between the points $(-4, 7, 6)$ and $(0, 3, -2)$ if both points are in three-dimensional space?

(A) –5

(B) –2

(C) 0

(D) 2

(E) 5

GO ON

10. S_n is a sequence such that for $n > 1$, $a_1 = 3$ and $a_n = 2a_{n-1} - 5$. What is the fourth term of S_n?

(A) –11

(B) –5

(C) –3

(D) 1

(E) 6

11. A tree fell over and is now leaning against the top of Mrs. Collini's house. The height of the house is 20 meters and the base of the tree is 35 meters from the base of the house. What angle does the fallen tree make with the ground, rounded to the nearest degree?

(A) 25°

(B) 30°

(C) 35°

(D) 60°

(E) 65°

12. In the figure, *a* equals

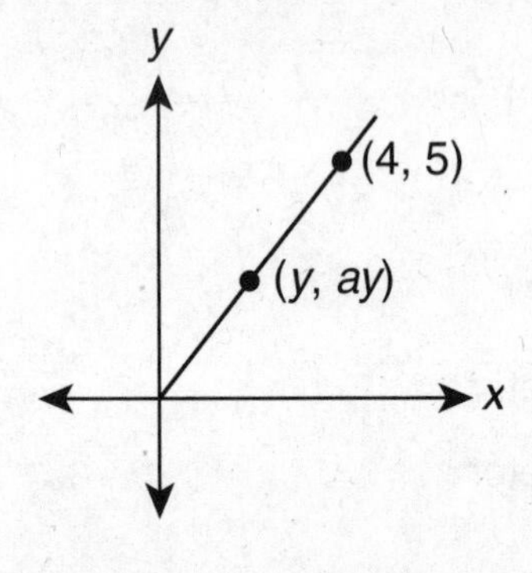

(A) $\frac{1}{5}$

(B) $\frac{1}{4}$

(C) $\frac{4}{5}$

(D) $\frac{5}{4}$

(E) 5

13. Which of the following is the solution, written in interval notation, to the inequality $3x^2 + x > 2x^2 + 4x + 4$?

(A) $(-\infty, -1)$

(B) $(-\infty, -1) \cup (4, \infty)$

(C) $(-1, 4)$

(D) $(4, \infty)$

(E) All real numbers

GO ON

Name ______________________ Date __________ Class __________

SAT Subject Test Practice Test I: Math Level IIC *continued*

USE THIS SPACE FOR SCRATCH WORK.

14. If $f(x) = \tan^{-1}x$, $g(x) = 0.9x$, and x is in radians, for how many values of x does $f(x) = g(x)$?

(A) 0

(B) 1

(C) 3

(D) 4

(E) Infinitely many

15. The figure shown is a dart board made up of eight rectangles that are all the same size. What is the probability that a dart is thrown and lands in a rectangle containing a 4, given that the dart landed in a rectangle containing an even number.

1	2	3	4
4	1	2	2

(A) $\frac{1}{4}$

(B) $\frac{3}{8}$

(C) $\frac{2}{5}$

(D) $\frac{1}{2}$

(E) $\frac{5}{8}$

16. The number of mold spores, S, in a certain culture is given by the equation $S = 100e^{0.5t}$, where t is the number of days after 12:00 A.M., January 1. During what day in January does the number of spores equal 1 thousand?

(A) 2nd

(B) 4th

(C) 5th

(D) 8th

(E) 12th

GO ON

Name ______________________ Date __________ Class __________

SAT Subject Test Practice Test I: Math Level IIC *continued*

USE THIS SPACE FOR SCRATCH WORK.

17. A car is traveling on a straight road. Its velocity versus time is shown on the graph. Which of the following statements is true?

(A) The car is moving the fastest at $t = 0$.

(B) The car is moving the fastest at $t = 3$.

(C) The car is stopped at $t = 3$.

(D) The car is moving backwards after $t = 3$.

(E) The car is in the same position at $t = 0$ and $t = 6$.

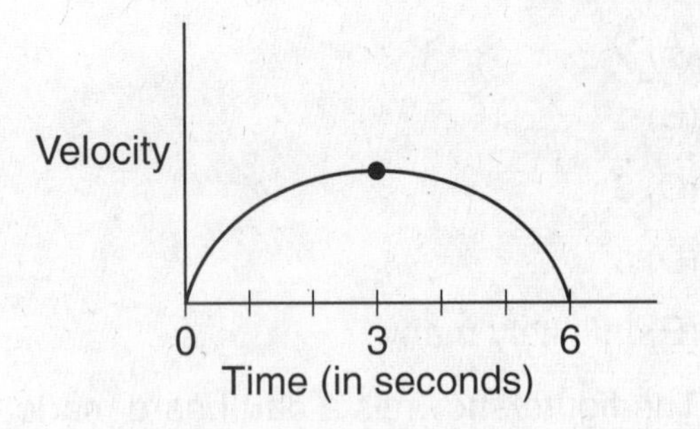

18. If $49^x = 7$ and $4^{2x+y} = \frac{1}{16}$, then $y =$

(A) -3

(B) -2

(C) -1

(D) 2

(E) 4

19. All license plate numbers in a certain state are composed of 3 letters followed by 3 digits. If the letters *I* and *O* cannot be used, which method would be used to find the number of different license plates possible for that state?

(A) $(24 + 10)^6$

(B) $(24 \cdot 10)^6$

(C) $24^3 \cdot 10^3$

(D) ${}_{24}P_3$

(E) ${}_{24}C_3$

20. The stem and leaf plot shows the heights, in inches, of 20 randomly chosen students in a large high school. What percent of the students are taller than 6 feet?

(A) 40

(B) 60

(C) 70

(D) 80

(E) 90

Stem	Leaf
5	8 9
6	0 2 3 4 5 5 7 8 9
7	1 3 3 4 7 9 9
8	0 1

GO ON

SAT Subject Test Practice Test I: Math Level IIC *continued*

USE THIS SPACE FOR SCRATCH WORK.

21. If (x_1, y_1) and (x_2, y_2) are the points of intersection of the circle whose equation is $x^2 + y^2 = 4$ and the line $y = x$, what is the value of $x_1 + x_2$?

(A) -1.414

(B) -0.707

(C) 0

(D) 0.707

(E) 1.414

22. If the domain of $f(x) = \sqrt{x^2 + 2x + 2c}$ is $(-\infty, -5] \cup [3, \infty)$, then $c =$

(A) 30

(B) 15

(C) 7.5

(D) -7.5

(E) -15

23. The graph of the parametric equation $\begin{cases} x = \sin t \\ y = \cos t \end{cases}$ is

(A) a line

(B) a parabola

(C) a hyperbola

(D) an ellipse

(E) a circle

24. If $5x - 6 = 3(y - 2)$, then $5\left(\frac{x}{y}\right)$ is

(A) -5

(B) 3

(C) 5

(D) 15

(E) 25

GO ON

Name ________________________ Date __________ Class __________

SAT Subject Test Practice Test I: Math Level IIC *continued*

25. If $\sin\left(x - \frac{\pi}{2}\right) = 0.2$, then $\cos x =$

USE THIS SPACE FOR SCRATCH WORK.

(A) $0.2 - \frac{\pi}{2}$

(B) $0.2 + \frac{\pi}{2}$

(C) $-0.2 + \frac{\pi}{2}$

(D) 0.2

(E) -0.2

26. What is the ratio of x to a if $x > a > 0$ and $\log(x + a) = 1 + \log(x - a)$?

(A) $\frac{12}{11}$

(B) $\frac{11}{10}$

(C) $\frac{11}{9}$

(D) $\frac{13}{10}$

(E) $\frac{10}{7}$

27. The polar coordinate of point B is (5, 60°). Approximately how many units above the x-axis is point B?

(A) 2.5

(B) 4.3

(C) 5

(D) 30

(E) 60

28. What is the range, in interval notation, of the piecewise function?

$$g(x) = \begin{cases} -3x + 5, & \text{if } -4 \leq x \leq 0 \\ 3x + 6, & \text{if } 0 < x \leq 4 \end{cases}$$

(A) $(0, \infty)$

(B) (5, 17]

(C) [5, 18]

(D) [6, 17]

(E) [6, 18]

GO ON

Name ______________________________ Date ____________ Class ____________

SAT Subject Test Practice Test I: Math Level IIC *continued*

USE THIS SPACE FOR SCRATCH WORK.

29. If $\sin(ax) = 0.3\cos(ax)$, x is in radians, $-\frac{\pi}{2} < x < \frac{\pi}{2}$, and $a > 1$, then $x =$

(A) $\frac{0.29}{a}$

(B) $\frac{0.30}{a}$

(C) $\frac{0.31}{a}$

(D) $\frac{0.96}{a}$

(E) $\frac{1.27}{a}$

30. If $A = \langle 2, 3 \rangle$ and $B = \langle 5, -1 \rangle$ are vectors, what is the length of vector C if $C = A - B$?

(A) –7

(B) 1

(C) 3

(D) 5

(E) 7

31. If $g(x) = 2x + 5$ and $g(f(x)) = x$, then $f(-3) =$

(A) −5

(B) −4

(C) −3

(D) −2

(E) −1

32. What is the approximate length of segment AB in the diagram?

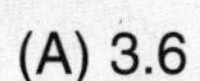

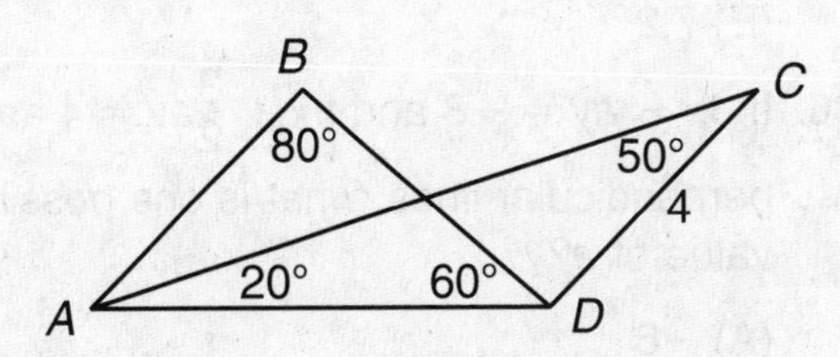

(A) 3.6

(B) 3.8

(C) 5.5

(D) 7.5

(E) 7.9

GO ON

SAT Subject Test Practice Test I: Math Level IIC *continued*

USE THIS SPACE FOR SCRATCH WORK.

33. If $f(x) = 10^{10x}$ and $g(x) = \log x$, then $g(g(f(a))) =$

(A) a

(B) e^a

(C) 10^a

(D) $\log a$

(E) $1 + \log a$

34. Line A has the equation $2x + 3y = 7$. If line B is perpendicular to line A at $x = 2$, where does line B intersect the x-axis?

(A) $-\frac{7}{2}$

(B) -7

(C) $\frac{4}{3}$

(D) $\frac{7}{3}$

(E) $\frac{7}{2}$

35. Let $f(x)$ be the equation of the line-of-best-fit used to approximate the data given in the chart. What is the approximate value of $f(5)$?

x	***y***
−3	0
−2	1
−1	1.5
0	1
1	3
2	4

(A) 5.0

(B) 5.8

(C) 6.4

(D) 7.0

(E) 7.2

36. If $3x - ay = -8$ and $8x + \frac{3}{2}ay = 4$ are perpendicular lines, what is one possible value of a?

(A) -6

(B) -2

(C) 4

(D) 6

(E) 8

GO ON

USE THIS SPACE FOR SCRATCH WORK.

37. The function $f(x)$ has the following values: $f(1) = 2$, $f(2) = 5$, $f(5) = 8$, and $f(8) = 10$. If $f^{-1}(x)$ is the inverse of $f(x)$, then $f^{-1}(5) =$

(A) 1

(B) 2

(C) 5

(D) 8

(E) 10

38. A 30-foot ladder rests against the top of a house that is 28 feet tall. A man stands vertically on the ladder. To the nearest degree, what angle does his body make with the ladder?

(A) 70°

(B) 68°

(C) 30°

(D) 21°

(E) 16°

39. If the line segment shown is rotated about the y-axis, it generates a solid with a volume of

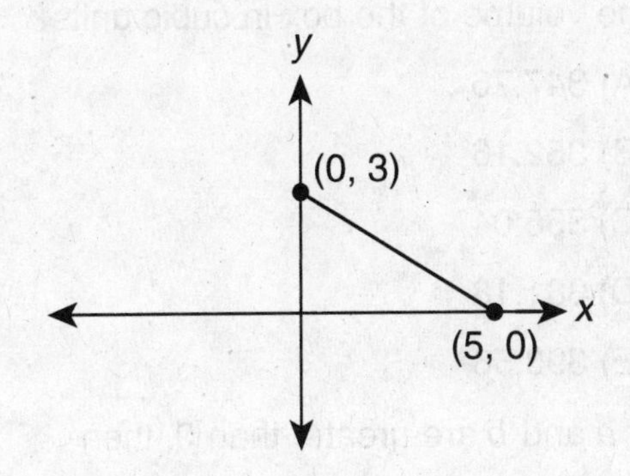

(A) 15π

(B) 25π

(C) 45π

(D) 75π

(E) 125π

40. The parametric forms of two curves are given. What is the approximate y-value of one of the points of intersection between the curves?

Curve 1: $x = 2t$, $y = t - 1$

Curve 2: $x = t + 3$, $y = t^2$

(A) 0.170

(B) 0.244

(C) 0.250

(D) 0.994

(E) 1.128

GO ON

SAT Subject Test Practice Test I: Math Level IIC *continued*

USE THIS SPACE FOR SCRATCH WORK.

41. Paul has a piggy bank where he keeps all his money. At the beginning of the year, he had x dollars. During the first month, he took 90% of the money out of the bank and then added \$100. During the second month, he took \$12 out and then took out another 40% of what was left. During the third month, he added \$30. If, at the end of the third month, the piggy bank contained \$90, how much money was in it at the beginning of the year?

(A) \$85

(B) \$90

(C) \$100

(D) \$115

(E) \$120

42. The length, width, and height of a box form an arithmetic sequence. If the surface area of the box is 365.04 square units and the box's smallest dimension is 3 units, what is the volume of the box in cubic units?

(A) 347.76

(B) 352.16

(C) 365.04

(D) 381.18

(E) 393.56

43. If a and b are greater than 0, then

$\sin(\arctan \frac{a}{b}) =$

(A) $\dfrac{b}{\sqrt{a^2 - b^2}}$

(B) $\dfrac{b}{\sqrt{b^2 - a^2}}$

(C) $\dfrac{b}{\sqrt{a^2 + b^2}}$

(D) $\dfrac{a}{\sqrt{a^2 - b^2}}$

(E) $\dfrac{a}{\sqrt{a^2 + b^2}}$

GO ON

Name ______________________ Date __________ Class __________

SAT Subject Test Practice Test I: Math Level IIC *continued*

USE THIS SPACE FOR SCRATCH WORK.

44. Find the value of a so that the matrix operation is correct.

$$\begin{bmatrix} a & b & c \\ 1 & 3 & -4 \\ 2 & 7 & -1 \end{bmatrix} \cdot \begin{bmatrix} 3 & 2 & 4 \\ -6 & 7 & 2 \\ 2 & 1 & 9 \end{bmatrix} = \begin{bmatrix} -11 & 13 & 37 \\ -23 & 19 & -26 \\ -38 & 52 & 13 \end{bmatrix}$$

(A) -5

(B) -3

(C) -2

(D) 2

(E) 5

45. What is the measure to the nearest degree of angle A shown in the figure?

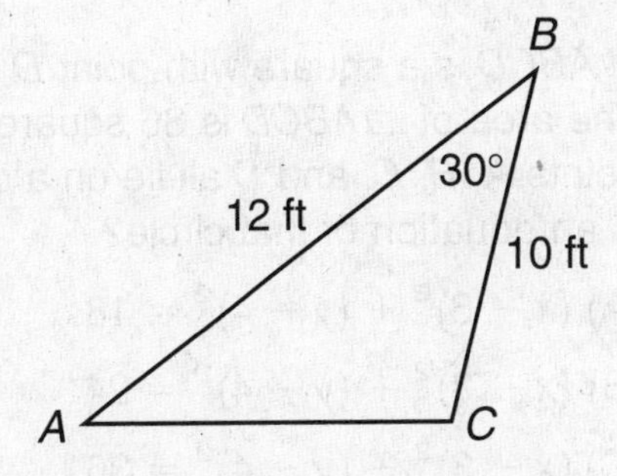

(A) 44°

(B) 56°

(C) 62°

(D) 70°

(E) 90°

46. If $f(x) = -a\sin(bx + c) + d$ and $a, b, c, d > 0$, what is the range of $f(x)$ in interval notation?

(A) $[-1, -1]$

(B) $[-a, a]$

(C) $[-d, d]$

(D) $[d - a, d + a]$

(E) $\left[-\frac{c}{b}, \frac{c}{b}\right]$

47. The cone shown has a height of 20 feet and a radius of 6 feet. If the cone is filled with water at a rate of 51.84π cubic feet per hour, what is the approximate height, in feet, of the water after 1 hour?

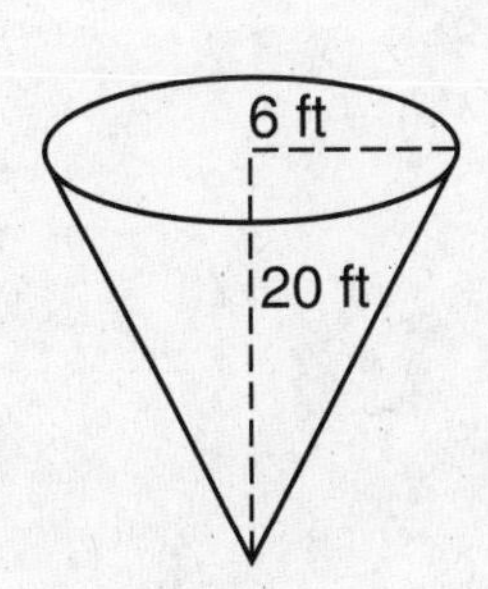

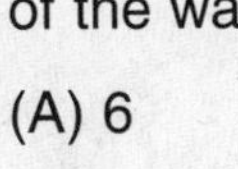

(A) 6

(B) 8

(C) 9

(D) 12

(E) 15

GO ON

Name ______________________ Date __________ Class __________

SAT Subject Test Practice Test I: Math Level IIC *continued*

USE THIS SPACE FOR SCRATCH WORK.

48. Solve the equation $be^{ax} \cdot e^{c} = 1$ for x.

(A) $\ln b^{-\frac{1}{a}} - \frac{c}{a}$

(B) $\frac{\ln\frac{1}{b} + c}{a}$

(C) $\frac{\ln b - c}{a}$

(D) $\ln\left(\frac{cb}{a}\right)$

(E) $\ln\left(\frac{c}{ab}\right)$

49. $\square ABCD$ is a square with point D at (3, 4). The area of $\square ABCD$ is 36 square units. If points A, B, C, and D all lie on a circle, what is an equation of that circle?

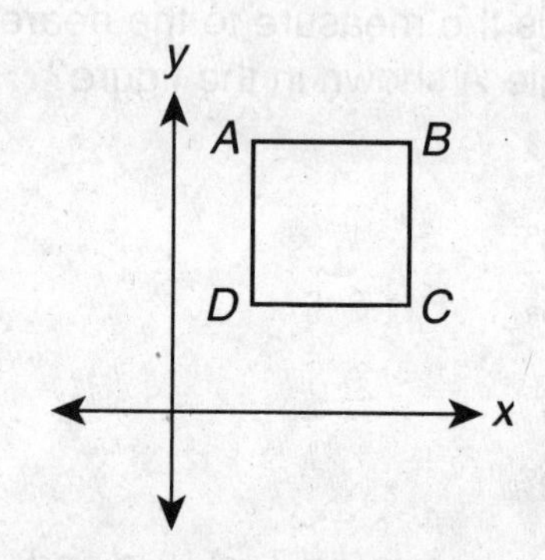

(A) $(x - 3)^2 + (y - 4)^2 = 18$

(B) $(x - 3)^2 + (y - 4)^2 = 24$

(C) $(x - 3)^2 + (y - 4)^2 = 36$

(D) $(x - 6)^2 + (y - 7)^2 = 18$

(E) $(x - 6)^2 + (y - 7)^2 = 36$

50. In the quadrilateral shown, $m\angle A = 90°$, $m\angle B = 90°$, $m\angle C = x^2 - 2x + 116$, and $m\angle D = 3x + 8$. What is (are) the value(s) of x?

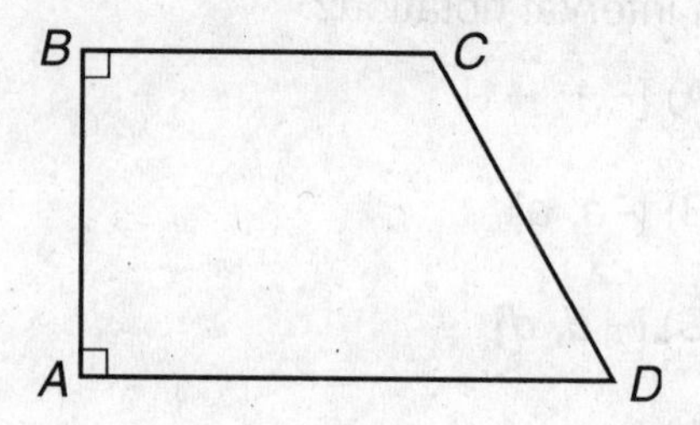

(A) −3

(B) −3 and 2

(C) 7

(D) 8

(E) 12

STOP **If you finish before time is called, you may check your work on this section only. Do not turn to any other section in the test.**

Name ______________________ Date __________ Class __________

SAT Subject Test Practice Test II: Math Level IIC

Time—60 minutes, 50 Questions

All questions in the Math Level 1 and Math Level 2 Tests are multiple-choice questions in which you are asked to choose the **BEST** response from the five choices offered. The directions for the tests are below:

Directions: For each of the following problems, decide which is the BEST of the choices given. If the exact numerical value is not one of the choices, select the choice that best approximates this value. Then fill in the corresponding oval on the answer sheet.

Notes:

1. A scientific or graphing calculator will be necessary for answering some (but not all) of the questions in this test. For each question you will have to decide whether or not you should use a calculator.
2. Level 1: The only angle measure used on this test is degree measure. Make sure your calculator is in the degree mode.
   Level 2: For some questions in this test you may have to decide whether your calculator should be in the radian mode or the degree mode.
3. Figures that accompany problems in this test are intended to provide information useful in solving the problems. They are drawn as accurately as possible EXCEPT when it is stated in a specific problem that its figure is not drawn to scale. All figures lie in a plane unless otherwise indicated.
4. Unless otherwise specified, the domain of any function f is assumed to be the set of all real numbers x for which $f(x)$ is a real number. The range of f is assumed to be the set of all real numbers $f(x)$, where x is in the domain of f.
5. Reference information that may be useful in answering the questions in this test can be found on the page preceding Question 1.

Reference Information. The following information is for your reference in answering some of the questions in this test.

Volume of a right circular cone with radius r and height h: $V = \frac{1}{3}\pi r^2 h$

Lateral Area of a right circular cone with circumference of the base c and slant height ℓ: $S = \frac{1}{2}c\ell$

Volume of a sphere with radius r: $V = \frac{4}{3}\pi r^3$

Surface Area of a sphere with radius r: $S = 4\pi r^2$

Volume of a pyramid with base area B and height h: $V = \frac{1}{3}Bh$

USE THIS SPACE FOR SCRATCH WORK.

1. What is an equation of the circle whose diameter has endpoints (1, 2) and (5, −6)?

   (A) $(x + 3)^2 + (y - 2)^2 = 20$

   (B) $(x - 3)^2 + (y + 2)^2 = 20$

   (C) $(x - 3)^2 - (y + 2)^2 = 20$

   (D) $(x - 3)^2 + (y + 2)^2 = 80$

   (E) $(x + 3)^2 - (y - 2)^2 = 80$

GO ON

SAT Subject Test Practice Test II: Math Level IIC *continued*

USE THIS SPACE FOR SCRATCH WORK.

2. A frequency distribution for the ages of 50 men in a certain association is given. What is the approximate mean age of the men?

Age	Frequency
32	4
35	20
36	11
37	5
38	7
40	3

(A) 35.7

(B) 35.9

(C) 36.3

(D) 36.5

(E) 37.1

3. What is the period of $f(x) = \sin(3x)\cos(4x)$?

(A) $\frac{\pi}{3}$

(B) $\frac{\pi}{2}$

(C) $\frac{2\pi}{3}$

(D) π

(E) 2π

4. A pyramid and a box share a base and have the same volume. What is the ratio of the height of the pyramid to the height of the box?

(A) 1 : 3

(B) 1 : 3

(C) 1 : 1

(D) 2 : 1

(E) 3 : 1

5. If a loan is compounded continuously at a rate of 5% per year, approximately how many years will it take for the loan amount to triple?

(A) 20

(B) 21

(C) 22

(D) 24

(E) 25

GO ON

Name ______________________ Date __________ Class __________

SAT Subject Test Practice Test II: Math Level IIC *continued*

USE THIS SPACE FOR SCRATCH WORK.

6. The graph of the polar function $r = 2\theta$ is a(n)

(A) spiral

(B) circle

(C) ellipse

(D) cardioid

(E) lemniscate

7. The area of circle $x^2 + y^2 + 6x + 4 = 0$ is

(A) 4π

(B) 5π

(C) 6π

(D) 8π

(E) 12π

8. Evaluate $\lim_{x \to 2^-} \frac{|x - 2|}{x - 2}$.

(A) Does not exist

(B) -2

(C) -1

(D) 1

(E) 2

9. Given the piecewise function

$$f(x) = \left\{ \begin{array}{ll} 3x, & \text{if } x < 0 \\ \tan x, & \text{if } 0 \le x \le \pi \\ x + 1, & \text{if } x > \pi \end{array} \right\},$$ where x is in radians, what are the values of x for which $f(x)$ is discontinuous?

(A) 0 and $\frac{\pi}{2}$

(B) 0, $\frac{\pi}{2}$, and π

(C) 0 and π

(D) $\frac{\pi}{2}$ and π

(E) $\frac{\pi}{2}$, π, and $\frac{3\pi}{2}$

GO ON

USE THIS SPACE FOR SCRATCH WORK.

10. If $(-3, y)$ is a point on the graph of the inverse of $f(x) = 2x^5 + 5x^3 + 6x + 10$, then $y = ?$

(A) -626

(B) -3

(C) -1

(D) 3

(E) 626

11. A triangle has sides of length 2, 3, and 4. What is the approximate measure of the smallest angle?

(A) 26°

(B) 27°

(C) 28°

(D) 29°

(E) 30°

12. The graph of $h(x)$ is a line. If $h(1) = 3$ and $h(5) = 9$, then an equation of $h(x)$ is

(A) $h(x) = -\frac{2}{3}x + \frac{5}{3}$

(B) $h(x) = \frac{2}{3}x + \frac{1}{3}$

(C) $h(x) = \frac{2}{3}x + \frac{5}{3}$

(D) $h(x) = -\frac{3}{2}x + \frac{3}{2}$

(E) $h(x) = \frac{3}{2}x + \frac{3}{2}$

13. Which of the labeled sections represent $(A \cup B) \cap C$

(A) I, II, III, IV, V, VI, and VII

(B) I, II, III, IV, V, and VI

(C) II, IV, V, VI, and VII

(D) II, IV, V, and VI

(E) IV, V, and VI

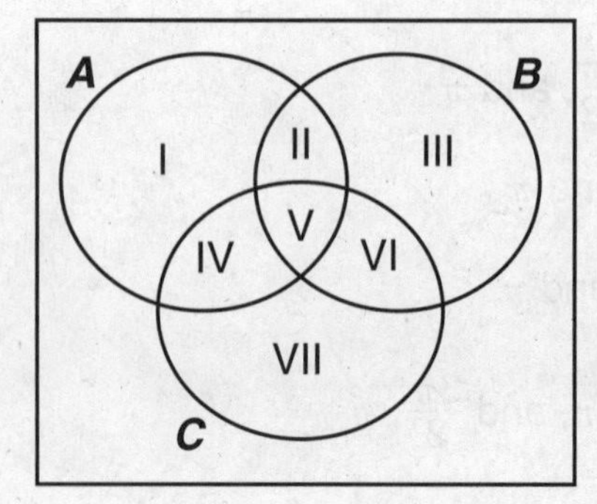

Name ______________________ Date __________ Class __________

SAT Subject Test Practice Test II: Math Level IIC *continued*

USE THIS SPACE FOR SCRATCH WORK.

14. The graph of $g(x) = \dfrac{(x+2)(x+3)(x+4)}{(x+2)(x-3)}$ has the following asymptotes.

(A) no horizontal or vertical asymptotes

(B) no horizontal and one vertical asymptote

(C) no horizontal and two vertical asymptotes

(D) one horizontal and one vertical asymptote

(E) one horizontal and two vertical asymptotes

15. Evaluate $\tan\left(\cos^{-1}\left(\frac{3}{8}\right)\right)$.

(A) $\frac{3\sqrt{55}}{55}$

(B) $\frac{8\sqrt{55}}{55}$

(C) $\frac{\sqrt{55}}{8}$

(D) $\frac{\sqrt{55}}{3}$

(E) $\frac{\sqrt{73}}{3}$

16. If $f(x) = x^2 + 2x$ then $\dfrac{f(x+h) - f(x)}{h} =$

(A) 1

(B) $2x + 2$

(C) $2x + 2 + h$

(D) $1 + \frac{4x}{h}$

(E) $2x + 2 + h + \frac{4x}{h}$

17. What is the probability of rolling two regular 6-sided dice and getting a sum of 10 or higher?

(A) $\frac{1}{10}$

(B) $\frac{1}{6}$

(C) $\frac{7}{36}$

(D) $\frac{5}{18}$

(E) $\frac{1}{2}$

GO ON

SAT Subject Test Practice Test II: Math Level IIC *continued*

USE THIS SPACE FOR SCRATCH WORK.

18. If $9.264^y = 3.264^x$, then $x =$

(A) $0.35y$

(B) $0.53y$

(C) $1.88y$

(D) $2.84y$

(E) $3.00y$

19. A cone shares its base with a cylinder and is inscribed inside the cylinder. If the volume of the cone is 24 cubic units, what is the volume in cubic units of the cylinder?

(A) 8

(B) 24

(C) 48

(D) 72

(E) Cannot be determined

20. Evaluate $(2 + 2i\sqrt{3})^3$.

(A) -64

(B) $8 - 24\sqrt{3}$

(C) $8 + 24\sqrt{3}$

(D) $8 - 24i\sqrt{3}$

(E) $8 + 24i\sqrt{3}$

21. If the points $(-1, -2)$, $(1, 6)$, and $(2, 1)$ all lie on the graph of $f(x) = ax^2 + bx + c$, then respectively a, b, and $c =$

(A) -3, -4, and -5

(B) -3, -4, and 5

(C) -3, 4, and 5

(D) 3, -4, and 5

(E) 3, 5, and 5

GO ON

USE THIS SPACE FOR SCRATCH WORK.

22. The statement "If it is expensive, then it is a car" is true. Which of the following statement(s) is (are) also true?

I. If it is not expensive, then it is not a car.
II. If it is not a car, then it is not expensive.
III. If it is a car, then it is expensive.

(A) I

(B) II

(C) III

(D) I and II

(E) I, II, and III

23. If the median of the data 2, 6, 3, 6, 4, x is 4.5, then $x =$

(A) 1

(B) 4.5

(C) 5

(D) 6

(E) 7

24. Use the parametric equations $y = -10t + 4$ and $x = 5t + 1$ to find y in terms of x.

(A) $y = -50x - 6$

(B) $y = -10x + 4$

(C) $y = -2x + 6$

(D) $y = 2x + 4$

(E) $y = 10x - 6$

25. The variable x varies inversely with the cube of y and directly with the square of z. If y is tripled and z is quadrupled, then x is multiplied by

(A) $\frac{1}{432}$

(B) $\frac{8}{27}$

(C) $\frac{15}{27}$

(D) $\frac{16}{27}$

(E) 432

GO ON

SAT Subject Test Practice Test II: Math Level IIC *continued*

USE THIS SPACE FOR SCRATCH WORK.

26. What is the sum of the three smallest non-negative solutions to $x\cos(10x) = 0$?

(A) $\frac{\pi}{5}$

(B) $\frac{9\pi}{20}$

(C) $\frac{4\pi}{5}$

(D) 2π

(E) $\frac{9\pi}{2}$

27. If the graph of $x^2 + y^2 + 2x + 6y + a = 0$ is a point, then $a =$

(A) -10

(B) -1

(C) 1

(D) 9

(E) 10

28. If $f(x) = \ln x^a$, then $f(be^c) =$

(A) $c \cdot b^a$

(B) $b^{(ac)}$

(C) $(bc)^a$

(D) $ac + \ln(ba)$

(E) $a(c + \ln b)$

29. The operation @ is defined by $x @ y = y - \frac{x}{y}$.

For what value(s) of m does $3 @ m = 2$?

(A) -1 and 3

(B) 1

(C) 1 and 2

(D) 2

(E) 2 and 3

GO ON

Name ______________________ Date __________ Class __________

SAT Subject Test Practice Test II: Math Level IIC *continued*

USE THIS SPACE FOR SCRATCH WORK.

30. $\sum_{k=3}^{7} c =$

(A) c

(B) $2c$

(C) $3c$

(D) $4c$

(E) $5c$

31. The radius of the circle shown is 9 units. If the area of the shaded sector is 9π square units, what is the length of $\overset{\frown}{ACB}$?

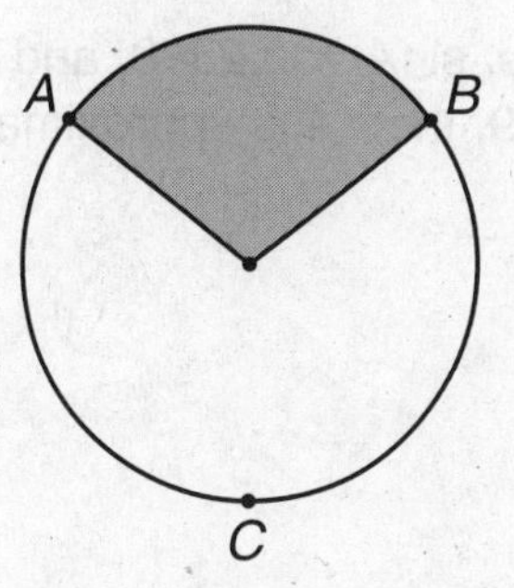

(A) 3π

(B) 15π

(C) 16π

(D) 18π

(E) 27π

32. The vertex of the graph of the parabola $x^2 + ax - y + b = 0$ is $\left(-\frac{5}{2}, -\frac{1}{4}\right)$. What is the sum of a and b?

(A) -2.75

(B) 2.25

(C) 5.25

(D) 7.75

(E) 11.00

33. The region bounded by the x-axis, the y-axis, $x = 2$, and $y = 3$ is revolved about the x-axis. What is the approximate volume of the solid generated?

(A) 37.7

(B) 43.5

(C) 47.1

(D) 51.3

(E) 56.5

GO ON

Name ______________________________ Date __________ Class __________

SAT Subject Test Practice Test II: Math Level IIC *continued*

USE THIS SPACE FOR SCRATCH WORK.

34. If the zeros of $f(x)$ are 1, 2, and 3, then the zeros of $f(x^2)$ are

(A) 1, 2, and 3

(B) ±1, ±4, and ±9

(C) 1, 4, and 9

(D) 1, $\sqrt{2}$, and $\sqrt{3}$

(E) ±1, $\pm\sqrt{2}$, and $\pm\sqrt{3}$

35. If A is in radians, $\sin A = 0.2588$, and $\tan A = -0.2679$, then A is approximately

(A) –0.26

(B) 0.26

(C) 2.88

(D) 6.02

(E) 6.28

36. If (a, b) is the point on the line $2x + 3y = 7$ that is closest to the origin, then $a =$

(A) $\frac{13}{12}$

(B) $\frac{14}{13}$

(C) $\frac{15}{14}$

(D) $\frac{16}{15}$

(E) $\frac{17}{16}$

37. If $f(x) = e^x$, then $\frac{f^{-1}(1)}{f(1)} =$

(A) 0

(B) 0.18

(C) 0.37

(D) 0.74

(E) 2.71

GO ON

SAT Subject Test Practice Test II: Math Level IIC *continued*

USE THIS SPACE FOR SCRATCH WORK.

38. If $x_3 = 11$ and $x_{n+1} = 2x_n + 1$, then $x_1 =$

(A) 2

(B) 5

(C) 9

(D) 23

(E) 47

39. If $\log_3 x = \log_2 x$, then $x =$

(A) $-\frac{3}{2}$

(B) -1

(C) 0

(D) 1

(E) 2

40. If $\sin(Ax) = \sin(2Ax)$, then which of the following could be the value of x?

I. $x = \frac{\pi}{3A}$

II. $x = \frac{\pi}{A}$

III. $x = \frac{2\pi}{A}$

(A) II

(B) III

(C) I and II

(D) II and III

(E) I, II, and III

41. If $c = e^{a+b}$, then which of the following is (are) true?

I. $a = \ln\left(\frac{c}{e^b}\right)$

II. $a = \ln c - b$

III. $a = \ln\left(\frac{c}{b}\right)$

(A) I

(B) II

(C) III

(D) I and II

(E) I, II, and III

GO ON

USE THIS SPACE FOR SCRATCH WORK.

42. If $f(x) = 3x^2 + 1$ and $g(x) = 2x + 3$, then $f(g(x)) = g(g(x))$ when $x =$

(A) -1.86 and -0.81

(B) -1.77 and -0.89

(C) -1.50 and -1.17

(D) -0.55 and 1.22

(E) At no values

43. A sphere is cut through its center vertically and half the sphere is discarded. The remaining semi-sphere is then cut horizontally through its original center and the bottom half is discarded. The remaining piece is then placed on a table and is cut vertically through the original center so that one of the pieces has a corner whose angle measures 30°. What is the ratio of the volume of the piece with the 30° angle to the volume of the original sphere?

(A) $\frac{1}{36}$

(B) $\frac{1}{24}$

(C) $\frac{1}{12}$

(D) $\frac{1}{6}$

(E) $\frac{1}{4}$

44. How many different ways can the letters in the word ERASER be arranged so that there is a vowel on each end?

(A) 36

(B) 72

(C) 144

(D) 180

(E) 720

GO ON

Name ______________________ Date __________ Class __________

SAT Subject Test Practice Test II: Math Level IIC *continued*

USE THIS SPACE FOR SCRATCH WORK.

45. For which of the following is $\dfrac{\frac{x}{y}}{\frac{x}{y} - \frac{y}{x}} > 0$?

(A) $x < y$

(B) $y > x$

(C) $x < y < 0$

(D) No values of x and y

(E) All values of x and y

46. Which of the following is part of the domain written in interval notation of $f(x) = \log\left(\cos\left(\frac{ax}{b} + \pi\right)\right)$?

(A) $\left(-\frac{2b\pi}{a}, -\frac{b\pi}{a}\right)$

(B) $\left(-\frac{2a\pi}{b}, 0\right)$

(C) $\left(-\frac{3b\pi}{2a}, -\frac{b\pi}{2a}\right)$

(D) $\left(-\frac{a\pi}{b}, -\frac{a\pi}{2b}\right)$

(E) $\left(-\frac{b\pi}{a}, 0\right)$

47. The expression $\dfrac{\tan^2 x + 1}{\sec^2 x - 1}$ is equivalent to

(A) $\sin^2 x$

(B) $\cos^2 x$

(C) $\tan^2 x$

(D) $\csc^2 x$

(E) $\sec^2 x$

48. If $f(x) = x^2$, $g(x) = \sin x$, $h(x) = \cos x$, and $j(x) = 2x + 1$, then which of the following is an even function?

(A) $h(j(x))$

(B) $f(j(x))$

(C) $g(h(x))$

(D) $g(j(x))$

(E) $j(g(x))$

GO ON

SAT Subject Test Practice Test II: Math Level IIC *continued*

USE THIS SPACE FOR SCRATCH WORK.

49. If the graph of
$2x^2 + 6y^2 + 4x + 12y - 46 = 0$
is inscribed inside a rectangle, the area of the rectangle is

(A) $18\sqrt{2}$

(B) $18\sqrt{3}$

(C) 36

(D) $27\sqrt{3}$

(E) $36\sqrt{3}$

50. What is the amplitude of the graph of
$y = 3\sin(ax)\cos(bx) + 3\sin(bx)\cos(ax)$?

(A) 3

(B) 6

(C) $3ab$

(D) $3(a + b)$

(E) $6(a + b)$

STOP **If you finish before time is called, you may check your work on this section only. Do not turn to any other section in the test.**

Name ______________________ Date ____________ Class ____________

ACT Assessment *Answer Sheet*

SIDE 1

USE A SOFT LEAD PENCIL ONLY.

A **NAME, ADDRESS, AND TELEPHONE (Please Print)**

Last Name First Name MI (Middle Initial)

House Number and Street

City State ZIP Code

Area Code / Phone Number

All examinees MUST complete blocks A, B, C, and D.

Registered Examinees: Enter the MATCHING INFORMATION in the blocks B, C, and D EXACTLY as it appears on your admission ticket, even if any part of the information is missing or incorrect. Fill in the corresponding ovals. If you do not complete these blocks to match your admission ticket EXACTLY, your score will be delayed. Leave block E blank.

Standby Examinees: Enter your identifying information in blocks B, C, and D. Fill in the corresponding ovals. Also fill in the Standby Testing oval in block E..

B **MATCH NAME** (First 5 letters of Last Name)

A B C D E F G H I J K L M N O P Q R S T U V W X Y Z

C **SOCIAL SECURITY NUMBER** (or ACT ID Number, including the leading dash)

– –

1 2 3 4 5 6 7 8 9 0

D **DATE OF BIRTH**

Month: Jan. Feb. Mar. Apr. May June July Aug. Sep. Oct. Nov. Dec.

Day: 1 2 3 4 5 6 7 8 9 0

Year: 1 2 3 4 5 6 7 8 9 0

Name ______________________ Date __________ Class __________

ACT Assessment *Answer Sheet*

SIDE 2

Test Directions

Mark only one oval for each question. Fill in your response completely and neatly. Erase any errors thoroughly.

Incorrect Marks:

Correct Mark:

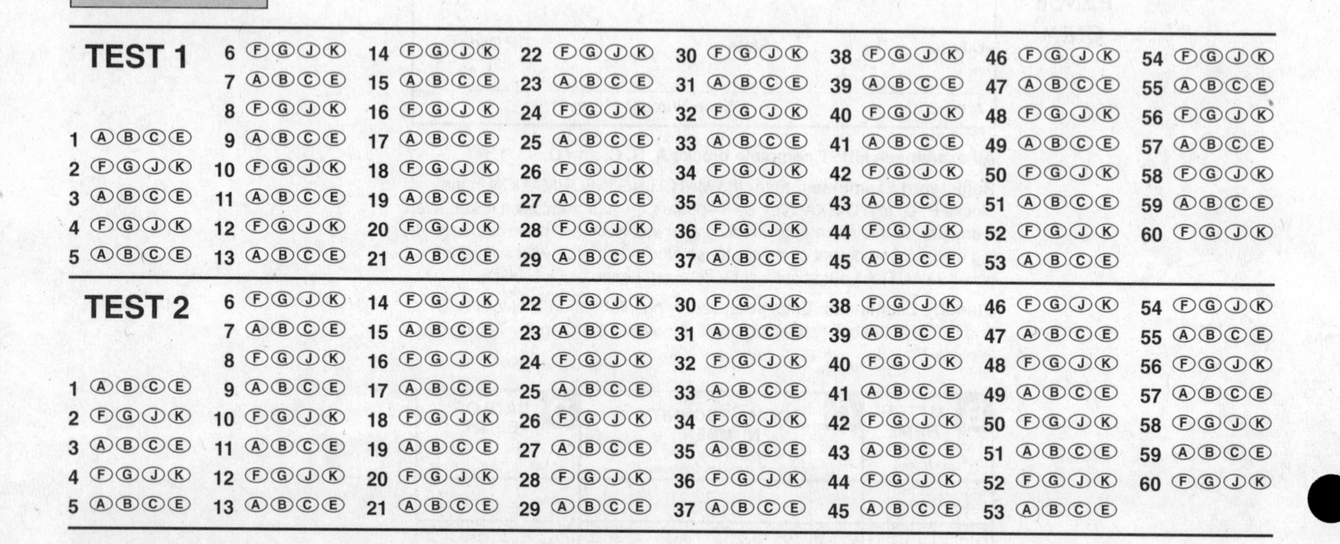

TEST 1

	6 F G J K	14 F G J K	22 F G J K	30 F G J K	38 F G J K	46 F G J K	54 F G J K
	7 A B C E	15 A B C E	23 A B C E	31 A B C E	39 A B C E	47 A B C E	55 A B C E
	8 F G J K	16 F G J K	24 F G J K	32 F G J K	40 F G J K	48 F G J K	56 F G J K
1 A B C E	9 A B C E	17 A B C E	25 A B C E	33 A B C E	41 A B C E	49 A B C E	57 A B C E
2 F G J K	10 F G J K	18 F G J K	26 F G J K	34 F G J K	42 F G J K	50 F G J K	58 F G J K
3 A B C E	11 A B C E	19 A B C E	27 A B C E	35 A B C E	43 A B C E	51 A B C E	59 A B C E
4 F G J K	12 F G J K	20 F G J K	28 F G J K	36 F G J K	44 F G J K	52 F G J K	60 F G J K
5 A B C E	13 A B C E	21 A B C E	29 A B C E	37 A B C E	45 A B C E	53 A B C E	

TEST 2

	6 F G J K	14 F G J K	22 F G J K	30 F G J K	38 F G J K	46 F G J K	54 F G J K
	7 A B C E	15 A B C E	23 A B C E	31 A B C E	39 A B C E	47 A B C E	55 A B C E
	8 F G J K	16 F G J K	24 F G J K	32 F G J K	40 F G J K	48 F G J K	56 F G J K
1 A B C E	9 A B C E	17 A B C E	25 A B C E	33 A B C E	41 A B C E	49 A B C E	57 A B C E
2 F G J K	10 F G J K	18 F G J K	26 F G J K	34 F G J K	42 F G J K	50 F G J K	58 F G J K
3 A B C E	11 A B C E	19 A B C E	27 A B C E	35 A B C E	43 A B C E	51 A B C E	59 A B C E
4 F G J K	12 F G J K	20 F G J K	28 F G J K	36 F G J K	44 F G J K	52 F G J K	60 F G J K
5 A B C E	13 A B C E	21 A B C E	29 A B C E	37 A B C E	45 A B C E	53 A B C E	

EXAMINEE STATEMENT AND SIGNATURE
(Read and sign your name as you would an official document.)

I hereby agree to the conditions set forth in the ACT Assessment registration booklet or web site instructions for the exam including the arbitration and dispute remedy provisions. I certify that I am the person whose name and address appear on this form.

____
Your Signature

____
Today's Date

Name ______________________ Date __________ Class __________

ACT Practice Test 1 Section 1

Time—60 minutes, 60 Questions

DIRECTIONS: Solve each problem, choose the correct answer, and then fill in the corresponding oval on your answer document.

Do not linger over problems that take too much time. Solve as many as you can; then return to the others in the time you have left for the test.

You are permitted to use a calculator on this test. You may use your calculator for any problems you choose, but some of the problems may best be done without using a calculator.

DO YOUR FIGURING HERE.

Notes: Unless otherwise stated, all of the following should be assumed.

1. Illustrative figures are NOT necessarily drawn to scale.
2. Geometric figures lie in a plane.
3. The word *line* indicates a straight line.
4. The word *average* indicates arithmetic mean.

---

1. Ten thousand tires are being stored in a warehouse. Two percent of the tires are not usable. What is the ratio of usable tires to not usable tires?

A.  $\frac{1}{98}$

B. $\frac{1}{50}$

C. $\frac{1}{49}$

D. $\frac{49}{1}$

E. $\frac{50}{1}$

DO YOUR FIGURING HERE.

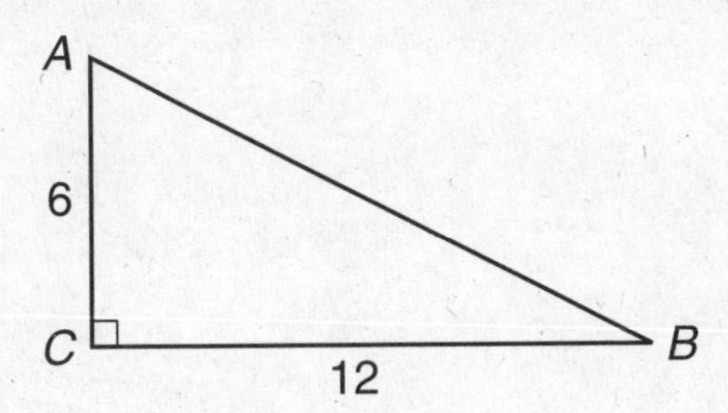

2. What is the perimeter of triangle ABC?

F. $6\sqrt{2}$

G. $6\sqrt{5}$

H. $18 + 6\sqrt{2}$

J. $18 + 6\sqrt{5}$

K. $24\sqrt{5}$

GO ON

Name ______________________ Date __________ Class __________

ACT Practice Test 1 Section 1 *continued*

DO YOUR FIGURING HERE.

3. How many real roots does the equation $x^2 + 9 = 0$ have?

A. 0

B. 1

C. 2

D. 3

E. 9

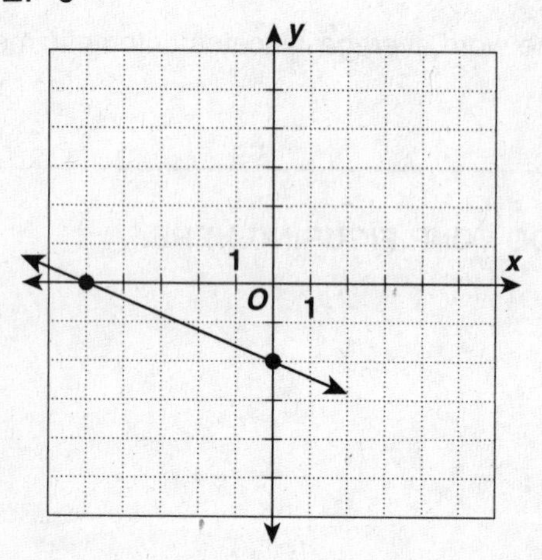

4. What is the slope of the graphed line?

F. $-\frac{5}{2}$

G. -2

H. $-\frac{2}{5}$

J. $\frac{2}{5}$

K. $\frac{5}{2}$

5. Every item on the clearance rack at a store is marked down 30%. Which of the following is a function to represent the price after the discount?

A. $f(x) = 0.3x$

B. $f(x) = 0.4x$

C. $f(x) = 0.7x$

D. $f(x) = 3.0x$

E. $f(x) = 7.0x$

GO ON

ACT Practice Test 1 Section 1 *continued*

DO YOUR FIGURING HERE.

6. Which of the following is equivalent to $(b)(b)(b)(b)(b) + (b)(b)(b)(b)(b)(b)$?

F. $11b$

G. 11^b

H. $b + 11$

J. $b^5 + b^6$

K. b^{11}

7. Which of the following is the best approximation of $\frac{\sqrt{145}}{\sqrt{17}}$?

A. 3

B. 12

C. 17

D. 22

E. 30

8. A farmer is plowing a rectangular field that is 400 feet long and 700 feet wide. If the farmer can plow approximately 500 square feet per minute, about how long will it take him to plow the whole field?

F. 1 hour

G. 2 hours

H. 4 hours

J. 9 hours

K. 20 hours

9. Simplify $2(6x + 7) - 5(x + 3)$.

A. $7x - 1$

B. $7x + 1$

C. $7x + 19$

D. $17x - 1$

E. $17x + 19$

GO ON

Name ______________________________ Date ____________ Class ____________

ACT Practice Test 1 Section 1 *continued*

DO YOUR FIGURING HERE.

10. There are 30 antique cars in a parade. Six of the cars are red, 14 are black, 5 are blue, and 5 are white. If a circle graph is used to represent this information, what percent of the graph would accurately represent the number of red cars?

F. 5

G. 6

H. 16.7

J. 20

K. 24

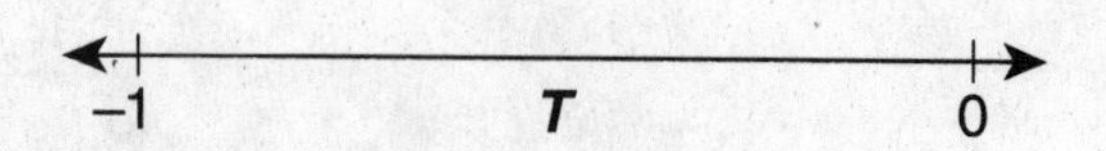

11. Which of the following values could NOT be the value of *T* on the number line above?

A. $-\frac{3}{5}$

B. $-0.\overline{6}$

C. $-\left(\frac{5}{7}\right)^2$

D. $-\left|-\frac{2}{3}\right|$

E. $\left(-\frac{4}{7}\right)^{-1}$

12. If $4y = 3x - 1$, then $3x =$

F. $4y + 1$

G. $\frac{4}{3}y - 1$

H. $\frac{4}{3}y + 1$

J. $\frac{4y - 1}{3}$

K. $\frac{4y + 1}{3}$

GO ON

Name ______________________ Date __________ Class __________

ACT Practice Test 1 Section 1 *continued*

DO YOUR FIGURING HERE.

13. The measure of the complement of the supplement of angle *A* is 28°. The measure in degrees of angle *A* is

A. 28

B. 45

C. 62

D. 90

E. 118

14. What is the image of point (−1, 4) under the translation (*x*, *y*) to (*x* + 1, *y* − 2)?

F. (−2, 6)

G. (−2, 2)

H. (−1, 2)

J. (0, 2)

K. (0, 6)

15. A cylindrical grain bin is being filled. The height of the grain bin is 20 feet and the diameter of its base is 10 feet. After the first 10 minutes, the height of the grain in the bin is 1 foot. At this rate, what will be the volume of the grain in the bin after the first hour?

A. 150 cubic feet

B. 100π cubic feet

C. 150π cubic feet

D. 600 cubic feet

E. 600π cubic feet

16. If $a = -2$ and $b = a^2$, then *a* and *b* are roots of which equation?

F. $x^2 + 2x - 8$

G. $x^2 - 2x - 8$

H. $x^2 + 2x + 8$

J. $x^2 - 6x - 8$

K. $x^2 - 6x + 8$

GO ON

ACT Practice Test 1 Section 1 *continued*

DO YOUR FIGURING HERE.

17. $\frac{3000}{10} + \frac{300}{100} + \frac{30}{1000} + \frac{3}{10000} = ?$

A. 300.303

B. 300.333

C. 303.0303

D. 303.33

E. 333.3

18. What is the center of a circle whose diameter has endpoints (5, 2) and (−1, −4)?

F. (2, −1)

G. (2, 0)

H. (3, −1)

J. (3, 0)

K. (1, 1)

19. If $x = \frac{1}{2}$ and $y = -6$, then $xy^2 =$

A. −18

B. −12

C. −6

D. 6

E. 18

20. Which statement best describes the relationship between the graphs of $y = 2$ and $x = 2$?

F. The two lines have the same slope.

G. The lines are perpendicular.

H. The lines are parallel.

J. The lines intersect at (2, 0).

K. None of the above.

GO ON

21. A new toy store is giving away 20 model airplanes; 9 are blue, 6 are red, and 5 are black. An airplane is selected at random and given to a customer. If the airplane is red, what is the probability that the next airplane, selected at random, is also red?

A. $\frac{5}{20}$

B. $\frac{6}{20}$

C. $\frac{5}{19}$

D. $\frac{6}{19}$

E. $\frac{14}{20}$

DO YOUR FIGURING HERE.

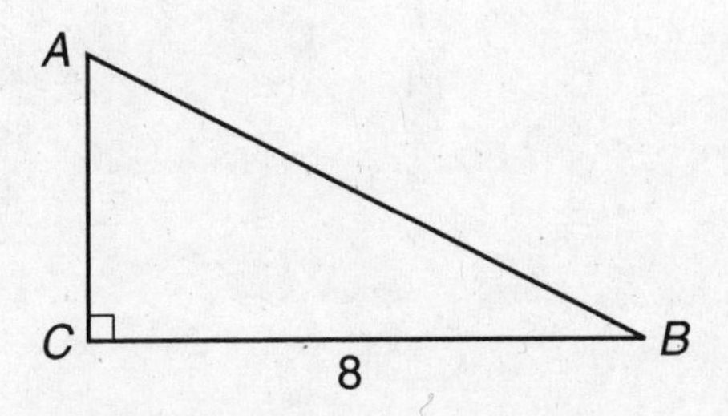

22. What is the length of side AB in $\triangle ABC$ shown above if $\cos B = \frac{2}{3}$?

F. 6

G. 8

H. 9

J. 10

K. 12

23. If $f(x) = 2x - 1$ and $g(x) = \sqrt{x + 5}$, what is $f(g(4))$?

A. 3

B. $\sqrt{12}$

C. 5

D. 15

E. 17

GO ON

ACT Practice Test 1 Section 1 *continued*

DO YOUR FIGURING HERE.

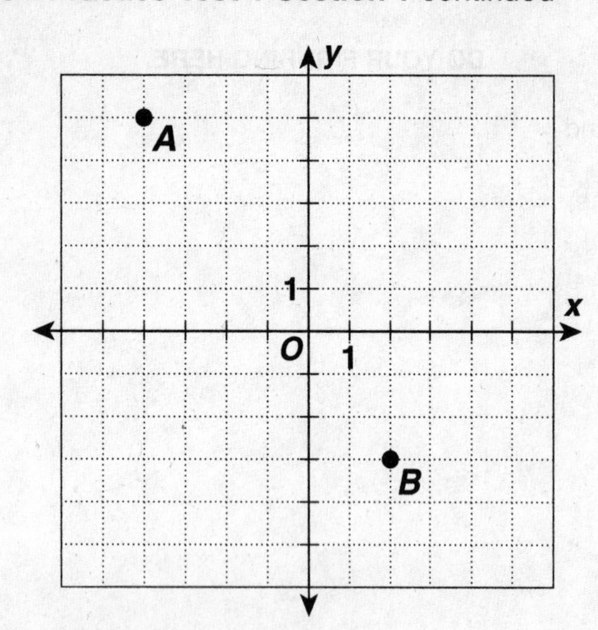

24. The distance between point A and point B shown in the coordinate grid is

F. 2

G. $2\sqrt{6}$

H. $2\sqrt{10}$

J. 10

K. $\sqrt{105}$

25. Which of the following is equal to $3\sqrt{60}$?

A. $2\sqrt{15}$

B. $6\sqrt{8}$

C. $5\sqrt{15}$

D. $6\sqrt{15}$

E. $12\sqrt{15}$

26. If $\begin{bmatrix} 2 & 1 \\ -6 & 0 \end{bmatrix} = \begin{bmatrix} 2 & b \\ 2a & 0 \end{bmatrix}$, what is the value of a?

F. −6

G. −3

H. −2

J. 2

K. 3

GO ON

Name ______________________ Date __________ Class __________

ACT Practice Test 1 Section 1 *continued*

DO YOUR FIGURING HERE.

27. The speed of light is 3×10^8 meters per second. If the sun is 1.5×10^{11} meters from Earth, and the distance from Pluto to the Sun is approximately 39.5 times the distance of Earth from the Sun, how many seconds does it take light to reach Pluto?

A. 1.975×10^2

B. 5.000×10^2

C. 7.595×10^2

D. 1.975×10^4

E. 7.595×10^4

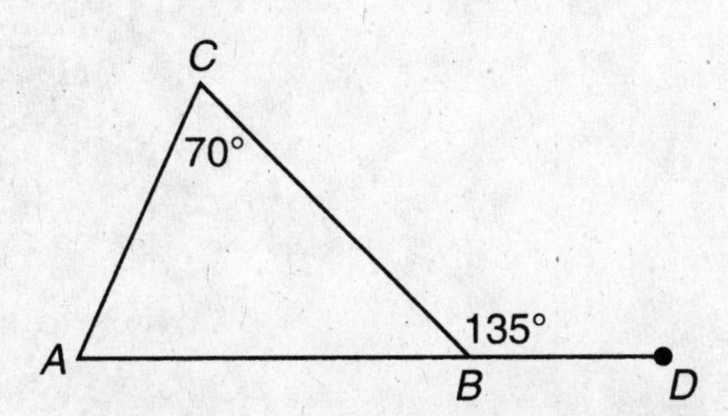

28. The diagram shows triangle *ABC* with segment *AB* extended to point *D*. The measure of angle *CBD* is 135° and the measure of angle *C* is 70°. What is the measure in degrees of angle *CAB*?

F. 45

G. 65

H. 70

J. 105

K. 110

29. Each figure in a pattern is a square whose width is one unit less than the width of the previous square. If the first square in the pattern has a perimeter of 40 units, what is the area in square units of the fifth square in the pattern?

A. 6

B. 24

C. 25

D. 36

E. 49

GO ON

ACT Practice Test 1 Section 1 *continued*

DO YOUR FIGURING HERE.

30. Which of the following is equivalent to $(3x^a)^b$?

F. $3x^{ab}$

G. $3x^{a+b}$

H. $3bx^{ab}$

J. $3^b x^{ab}$

K. $3^b x^{a+b}$

31. If $x + 3y = 1$ and $3x + y = 11$, then $3x + 3y = ?$

A. 3

B. 6

C. 9

D. 18

E. 36

32. Which of the following is the graph of the solution set of $-2x - 2 > x + 10$?

F. 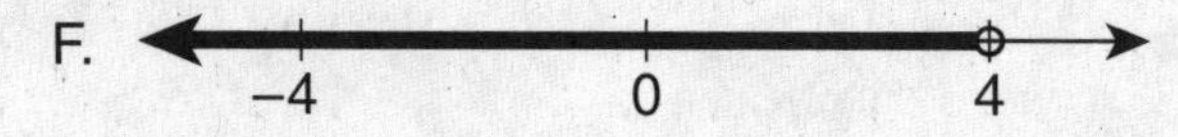

G. −4 0 4

H. −4 0 4

J. −4 0 4

K. −4 0 4

33. What is the circumference of a circle whose equation is $(x + 2)^2 + (y - 3)^2 = 36$?

A. 3π

B. 6π

C. 12π

D. 18π

E. 36π

GO ON

Name ______________________ Date __________ Class __________

ACT Practice Test 1 Section 1 *continued*

34. If A is a point in 3D-space with coordinates $(5, -1, 6)$, what is the approximate distance from the origin to point A?

F. 6

G. 8

H. 10

J. 13

K. 62

DO YOUR FIGURING HERE.

35. When the point $(-2, 3)$ is reflected across the x-axis, what are the coordinates of its image?

A. $(-2, -3)$

B. $(-3, 2)$

C. $(-2, 3)$

D. $(2, -3)$

E. $(3, -2)$

36. If m and n are factors of p, and $m = 6$ and $n = 9$, which of the following could NOT be the value of p?

F. 36

G. 48

H. 54

J. 72

K. 90

37. If $x = 3$, then $\dfrac{2}{\frac{x}{6} + \frac{6}{x}} = ?$

A. $\frac{2}{5}$

B. $\frac{1}{2}$

C. $\frac{4}{5}$

D. $\frac{5}{2}$

E. 5

GO ON

DO YOUR FIGURING HERE.

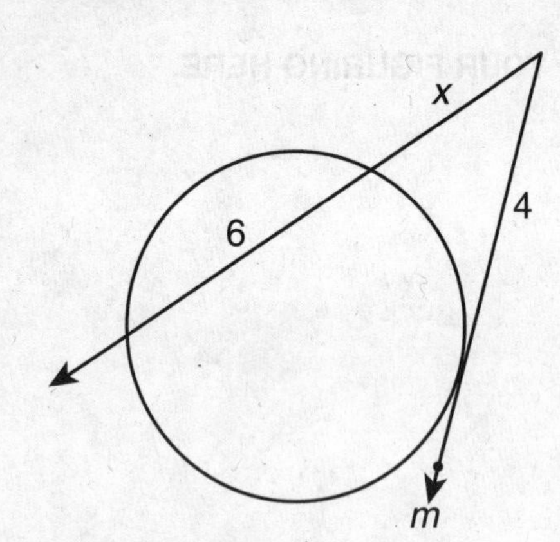

38. Line m is tangent to the circle above. What is the value of x?

F. -8

G. -2

H. 2

J. $\frac{8}{3}$

K. 8

39. What is the value of $\sin(2x + \pi)$ if $x = \frac{\pi}{4}$?

A. -1

B. $-\frac{1}{2}$

C. 0

D. $\frac{1}{2}$

E. 1

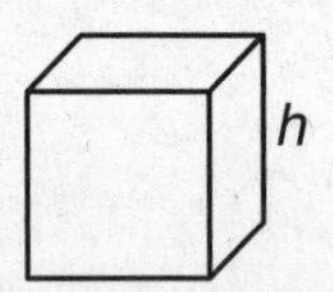

40. The solid above is a cube and the value of h is an integer. Which of the following could NOT be the volume of the cube?

F. 1

G. 3

H. 8

J. 27

K. 64

GO ON

Name ______________________ Date __________ Class __________

ACT Practice Test 1 Section 1 *continued*

DO YOUR FIGURING HERE.

41. A set of data has 10 values, no two of which are the same. If the smallest data value is removed from the set, which of the following statements MUST be true?

A. The range of the first data set is greater than the range of the second data set.

B. The mode of the first data set is greater than the mode of the second data set.

C. The medians of the two data sets are the same.

D. The mean of the first data set is greater than the mean of the second data set.

E. The maximum value of the first data set is greater than the maximum value of the second data set.

42. If x does not equal 0 and $\frac{2x^3 - 3x^2}{x^2} < 3$, then x could be any of the following EXCEPT:

F. -3

G. -1

H. 1

J. 2

K. 3

43. A high school baseball team has 4 pitchers, 2 catchers, 3 first basemen, and 1 person for every other position. No person plays more than one position. How many different configurations of players can the coach put on the field?

A. 9

B. 12

C. 24

D. 48

E. 108

GO ON

DO YOUR FIGURING HERE.

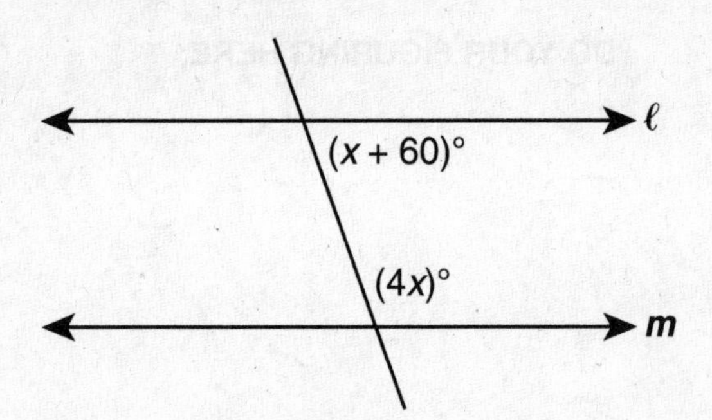

44. If line ℓ is parallel to line m above, what is the value of x?

F. 12

G. 15

H. 20

J. 24

K. 30

45. The maximum y-value of the graph of $y = 6\sin x - 1$ is

A. −1

B. 0

C. 1

D. 5

E. 6

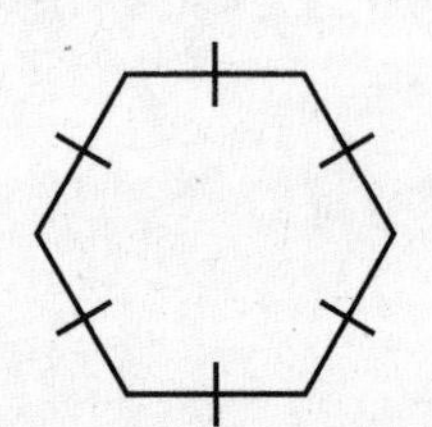

46. Which of the following statements is true about the polygon shown above?

F. The figure is an octagon.

G. The sum of the figure's interior angles is 540°.

H. The sum of the figure's exterior angles is 720°.

J. The measure of each of the figure's interior angles is 120°.

K. The figure has only one line of symmetry.

GO ON

DO YOUR FIGURING HERE.

47. For the equation $ax^2 + bx + c = 0$, if $4ac > b^2$, then the equation has

A. two real roots

B. two complex roots

C. one real root and one complex root

D. no roots

E. an infinite number of roots

48. The measures of two adjacent angles of a parallelogram are in the ratio 3 : 5. The measure in degrees of the smaller angle is

F. 22.5

G. 33.5

H. 67.5

J. 112.5

K. 135

49. Points *a*, *b*, *c*, and *d* all lie on the same line and $a < b < c < d$. If the distance from *a* to *b* is 3, the distance from *a* to *d* is 11, and the distance from *c* to *d* is 2, what is the distance from *b* to *d*?

A. 3

B. 4

C. 5

D. 6

E. 8

50. When a furniture store sells a floor model, it marks the retail price of the model down 30%. Every 30 days after that, the price is marked down an additional 20% until it is sold. The store decides to sell a floor model on March 15th. If the retail price of the item was $1,200 and the item is sold on June 2nd, what was the final selling price of the item?

F. $360.00

G. $430.08

H. $537.60

J. $720.00

K. $768.00

GO ON

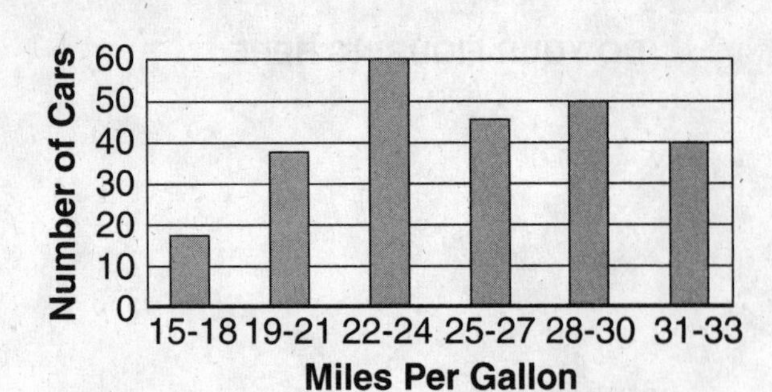

DO YOUR FIGURING HERE.

51. A car rental company monitored its fleet's gas mileage rates. The number of miles per gallon achieved by the company's 250 cars is depicted above. What percent of the cars achieved a gas mileage rating between 22 and 24 miles per gallon?

A. 24

B. 30

C. 42

D. 46

E. 60

52. What values of x satisfy the equation $|2x - 1| = |x + 4|$?

F. None

G. $x = 5$ only

H. $x = -\frac{5}{3}$ and $x = 5$

J. $x = -1$ and $x = 5$

K. $x = \frac{5}{3}$ and $x = 5$

53. If $x^2 + 3x - 10 > 0$, then x cannot be which of the following?

A. −7

B. −6

C. −3

D. 3

E. 6

GO ON

ACT Practice Test 1 Section 1 *continued*

DO YOUR FIGURING HERE.

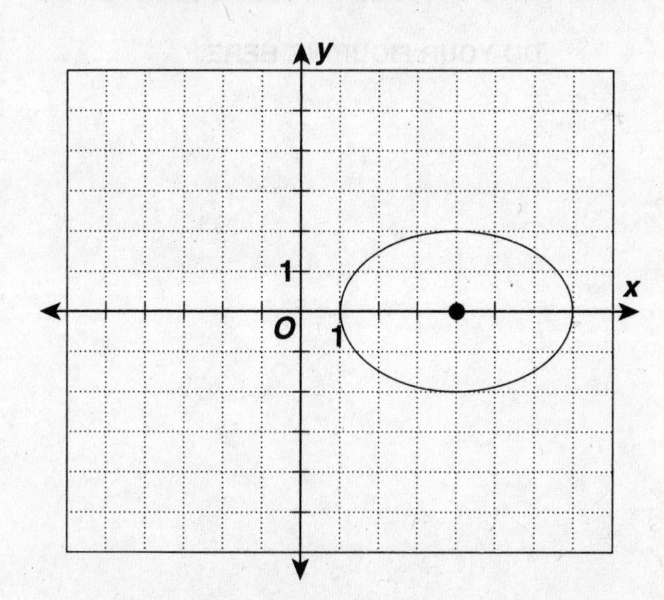

54. Which of the following could be the equation of the conic graphed above?

F. $\frac{(x-4)^2}{3} + \frac{y^2}{2} = 1$

G. $\frac{x^2}{3} + \frac{(y-4)^2}{2} = 1$

H. $\frac{(x-4)^2}{9} + \frac{y^2}{4} = 1$

J. $\frac{x^2}{9} + \frac{(y-4)^2}{4} = 1$

K. $\frac{(x-4)^2}{6} + \frac{y^2}{4} = 1$

55. What is the sum of the roots of the equation $(2x - 3)(3x + 1) = 0$?

A. -2

B. $-\frac{7}{6}$

C. $\frac{7}{6}$

D. $\frac{11}{6}$

E. 2

GO ON

Name ______________________ Date __________ Class __________

ACT Practice Test 1 Section 1 *continued*

DO YOUR FIGURING HERE.

x	−2	−1	1	5
$f(x)$	6	5	3	−1

56. The table of values given can be derived from which one of the following functions?

F. $f(x) = 2x + 10$

G. $f(x) = x + 2$

H. $f(x) = x^2 + 2$

J. $f(x) = 4 - x$

K. $f(x) = x - 6$

57. Which of the following values of x satisfies the equation $\cos 4x + 1 = 0$, if $\frac{\pi}{2} \leq x \leq \pi$?

A. $\frac{\pi}{4}$

B. $\frac{\pi}{2}$

C. $\frac{3\pi}{4}$

D. $\frac{5\pi}{6}$

E. $\frac{15\pi}{16}$

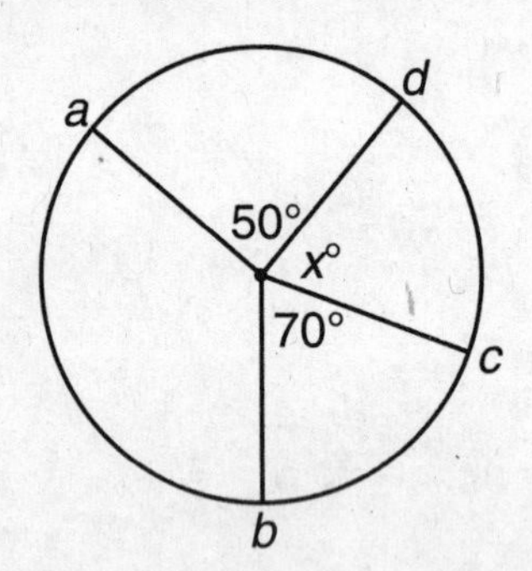

58. If the measure of $\widehat{ab}$ is equal to 2 times the measure of $\widehat{cd}$, what is the value of x in degrees?

F. 60

G. 80

H. 90

J. 140

K. 240

GO ON

Name ______________________ Date __________ Class __________

DO YOUR FIGURING HERE.

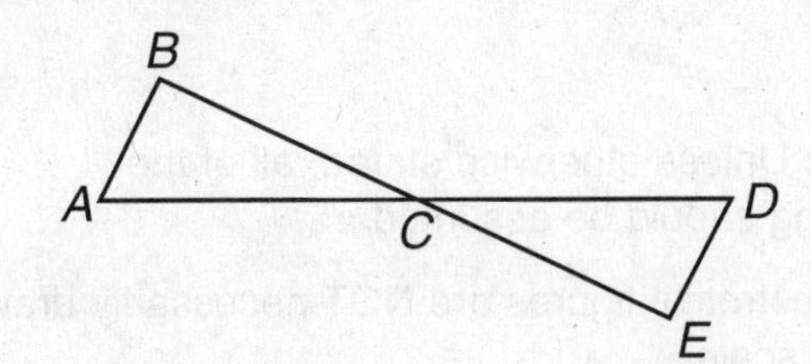

GIVEN: Point C is the midpoint of $\overline{AD}$ and $\overline{BE}$.
PROVE: $\triangle ABC \cong \triangle DEC$

Statement	Justification
1. C is the midpoint of $\overline{AD}$ and $\overline{BE}$.	Given
2. $\overline{AC} \cong \overline{DC}$ and $\overline{BC} \cong \overline{EC}$	Definition of midpoint
3. $\angle ACB \cong \angle DCE$	?
4. $\triangle ABC \cong \triangle DEC$	SAS

59. Which property provides justification for Statement 3 in the proof?

A. Reflexive Property

B. Transitive Property

C. Alternate interior angles are congruent.

D. Vertical angles are congruent.

E. Adjacent angles are congruent.

60. Simplify $\dfrac{1}{\frac{1}{2x} + \frac{x}{2}}$.

F. $\dfrac{2}{3x}$

G. $\dfrac{1}{x^2}$

H. $\dfrac{2x}{1 + x^2}$

J. $\dfrac{4x}{x + 1}$

K. $\dfrac{4x}{2 + x^2}$

STOP **If you finish before time is called, you may check your work on this section only. Do not turn to any other section in the test.**

Name ______________________ Date __________ Class __________

ACT Practice Test 2 Section 1

Time—60 minutes, 60 Questions

DIRECTIONS: Solve each problem, choose the correct answer, and then fill in the corresponding oval on your answer document.

Do not linger over problems that take too much time. Solve as many as you can; then return to the others in the time you have left for the test.

You are permitted to use a calculator on this test. You may use your calculator for any problems you choose, but some of the problems may best be done without using a calculator.

Notes: Unless otherwise stated, all of the following should be assumed.

1. Illustrative figures are NOT necessarily drawn to scale.
2. Geometric figures lie in a plane.
3. The word *line* indicates a straight line.
4. The word *average* indicates arithmetic mean.

DO YOUR FIGURING HERE.

1. A store buys an item and increases the price 80% before it is sold to the consumer. If the original price was \$40, how much does the consumer pay?

A. \$32

B. \$44

C. \$48

D. \$72

E. \$360

2. Which of the following is the best approximation of $\frac{3(11)^2}{\sqrt{143}}$?

F. 30

G. 65

H. 73

J. 91

K. 102

GO ON

Name ______________________________ Date __________ Class __________

ACT Practice Test 2 Section 1 *continued*

3. Mr. Smith has 40 students in his class. If 22 of the students are boys, what is the ratio of girls to boys?

DO YOUR FIGURING HERE.

A. $\frac{9}{40}$

B. $\frac{9}{20}$

C. $\frac{11}{20}$

D. $\frac{9}{11}$

E. $\frac{11}{9}$

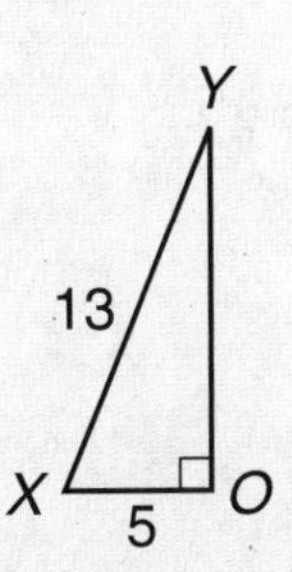

4. What is the area of triangle *XOY*?

F. 30

G. 32.5

H. 50.5

J. 60

K. 65

5. What is an equation of the line that passes through the origin and the point (4, 5)?

A. $y = \frac{4}{5}x$

B. $y = \frac{4}{5}x + 4$

C. $y = \frac{4}{5}x + 5$

D. $y = \frac{5}{4}x$

E. $y = \frac{5}{4}x + 5$

GO ON

ACT Practice Test 2 Section 1 *continued*

DO YOUR FIGURING HERE.

6. Which of the following is equivalent to $3(c)(c)(c)(c)(c)(c)(c) + 8(c)(c)(c)(c)(c)(c)(c)$?

F. $11c^7$

G. $24c^7$

H. $11c^{14}$

J. $24c^{14}$

K. $24 \cdot 7^c$

7. What is (are) the root(s) of the equation $x^2 + 16 = 0$?

A. 2

B. ± 2

C. 4

D. ± 4

E. $\pm 4i$

8. Pete is making a shirt out of cloth that costs $5 per square foot. If he buys a rectangular piece of cloth that is 2 feet by 4 feet, how much does he have to pay?

F. $5

G. $8

H. $20

J. $30

K. $40

9. Simplify $x(x + 4) + (x + 2)(x - 1)$.

A. $8x - 2$

B. $2x^2 + 2$

C. $2x^2 + 5x - 2$

D. $5x^3$

E. $7x^3 - 2$

GO ON

Name ______________________ Date __________ Class __________

ACT Practice Test 2 Section 1 *continued*

10. Which of the following is less than 0?

DO YOUR FIGURING HERE.

F. $\left(-\frac{1}{4}\right)^2$

G. $\left(\frac{1}{4}\right)^{-2}$

H. $\left(-\frac{1}{4}\right)^{-1}$

J. $\left|-\frac{1}{4}\right|$

K. $-\left(-\frac{1}{4}\right)$

11. There are 50 marbles in a bag. Twenty of the marbles are blue, 14 are red, and the rest are green. What percent of the marbles are green?

A. 16

B. 28

C. 32

D. 40

E. 50

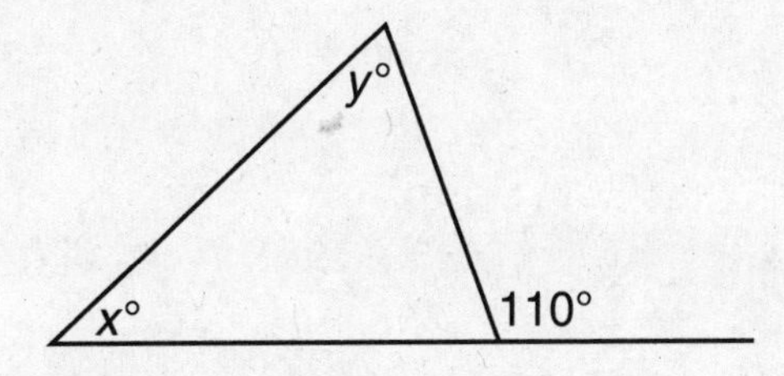

12. What is the sum of *x* and *y* in degrees?

F. 35

G. 40

H. 70

J. 110

K. 180

GO ON

ACT Practice Test 2 Section 1 *continued*

DO YOUR FIGURING HERE.

13. If $4x + 3 = 2x + y$, then $2x =$

A. $y - 6$

B. $y - 3$

C. $y + 3$

D. $2y - 3$

E. $2y + 3$

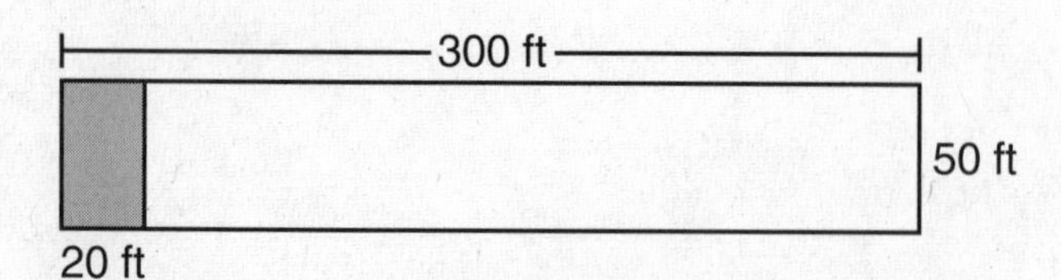

14. Billy is mowing the lawn shown above. The shaded region is the area of the lawn that Billy was able to mow in 15 minutes. At this rate, what is the area of the lawn that Billy will be able to mow in one hour?

F. 1000 ft^2

G. 2000 ft^2

H. 3000 ft^2

J. 4000 ft^2

K. 5000 ft^2

15. What is the image of the point (3, −2) under the translation (x, y) to $(x - 4, y - 1)$?

A. (−1, −3)

B. (7, −1)

C. (1, 1)

D. (1, −1)

E. (7, 1)

16. If $x = 8$ and $y = \frac{1}{4}$, then $-3x^2y^2 =$

F. −36

G. −18

H. −12

J. 18

K. 36

GO ON

Name ______________________ Date __________ Class __________

ACT Practice Test 2 Section 1 *continued*

DO YOUR FIGURING HERE.

17. Which of the following statements best describe the relationship between the graphs of $y = 2$ and $y = 2x + 5$?

A. The lines have the same slope.

B. The lines are perpendicular.

C. The lines intersect in exactly one point.

D. The lines intersect in more than one point.

E. None of the above.

18. If the diameter of a circle passes through the points $(-2, 3)$ and $(-5, 7)$, what is the radius of the circle?

F. 1.5

G. 2

H. 2.5

J. 4

K. 5

19. What is the product of the roots of the equation $x^2 + 7x + 10 = 0$?

A. -10

B. -7

C. $\frac{2}{5}$

D. 7

E. 10

20. $\left(\frac{1}{10}\right)^0 + \frac{1}{10} + \left(\frac{1}{10}\right)^2 + \left(\frac{1}{10}\right)^3 + \left(\frac{1}{10}\right)^4 =$

F. 0.1010101

G. 0.1111

H. 1.01010101

J. 1.1111

K. 10.1010101

GO ON

Name ______________________ Date ____________ Class ____________

ACT Practice Test 2 Section 1 *continued*

DO YOUR FIGURING HERE.

21. If $h(x) = \sqrt{3x + 6}$ and $m(x) = x^2 + 1$, what is $h(m(3))$?

A. 6

B. 7

C. 8

D. 9

E. 10

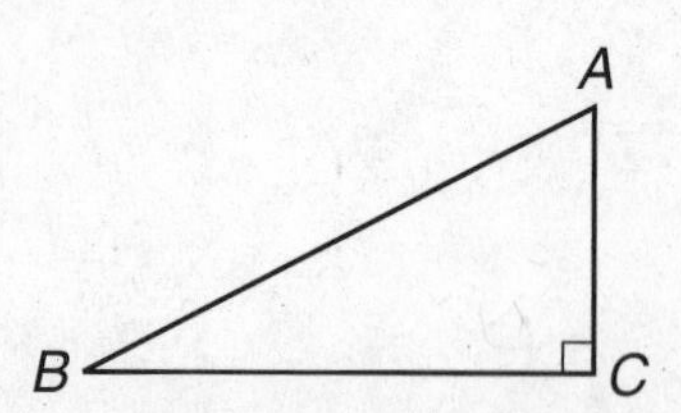

22. If $\tan A = \frac{5}{2}$, which of the following could be the hypotenuse of triangle ABC?

F. 2

G. 5

H. $\sqrt{29}$

J. 7

K. 10

23. Thirty students took a test. There were 10 A's, 15 B's, and 5 C's. If Mary earned an A, what is the probability that Pete earned a C?

A. $\frac{1}{9}$

B. $\frac{1}{6}$

C. $\frac{5}{29}$

D. $\frac{1}{5}$

E. $\frac{5}{9}$

GO ON

Name ______________________ Date __________ Class __________

ACT Practice Test 2 Section 1 *continued*

DO YOUR FIGURING HERE.

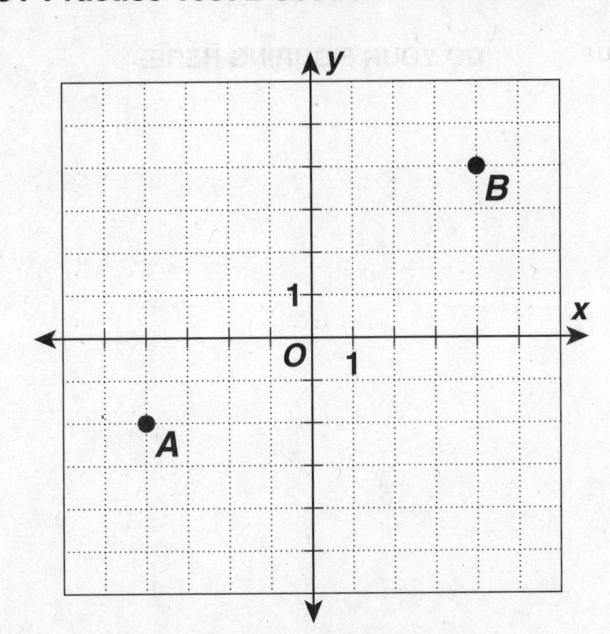

24. The midpoint between point *A* and point *B* on the coordinate grid is

F. (0, 0)

G. (0, 1)

H. (−1, 0)

J. (−1, 1)

K. (1, 0)

25. Each number in a list is five less than the number before it. If the first number is 145, what is the 7th number?

A. 100

B. 105

C. 110

D. 115

E. 120

26. Simplify $\frac{(9.2 \times 10^5)(4.6 \times 10^7)}{2 \times 10^8}$.

F. 2.116×10^4

G. 6.9×10^4

H. 2.116×10^5

J. 2.116×10^{20}

K. 6.9×10^{27}

GO ON

Name ______________________ Date __________ Class __________

ACT Practice Test 2 Section 1 *continued*

DO YOUR FIGURING HERE.

27. If $\begin{bmatrix} 1 & 2 \\ 3 & 4 \end{bmatrix} + \begin{bmatrix} 5 & 6 \\ 7 & 8 \end{bmatrix} = \begin{bmatrix} a & b \\ c & d \end{bmatrix}$, what is the value of b?

A. 6

B. 8

C. 10

D. 12

E. 14

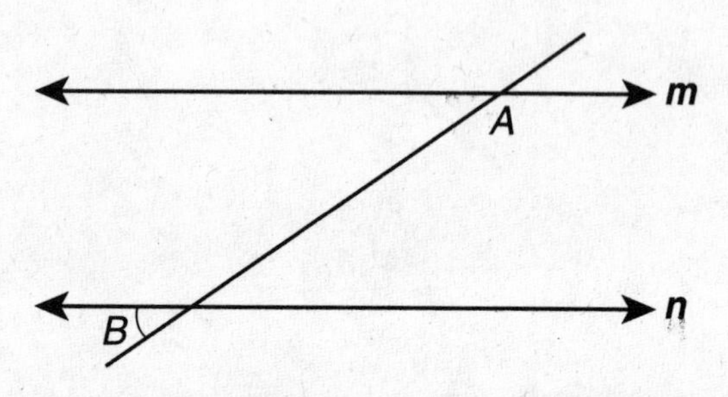

28. If lines m and n above are parallel and the measure of angle A is 110°, what is the measure of angle B in degrees?

F. 10

G. 70

H. 100

J. 170

K. 180

29. Which of the following is equal to $\sqrt{80}$?

A. $4\sqrt{5}$

B. $5\sqrt{4}$

C. $16\sqrt{5}$

D. $5\sqrt{16}$

E. $2\sqrt{40}$

GO ON

DO YOUR FIGURING HERE.

30. What is the area of a circle whose equation is $(x - 3)^2 + (y + 1)^2 = 81$?

F. 9

G. 18

H. 9π

J. 18π

K. 81π

31. If $5x + 3y = 7$ and $3x + 5y = 9$, then $2x - 2y = ?$

A. −4

B. −2

C. 0

D. 2

E. 4

32. Which of the following is the graph of the solution set of $x - 3 < 3x + 5$?

F. 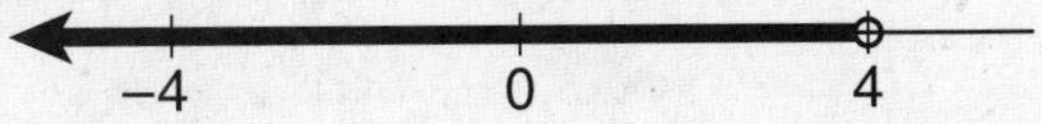

G. 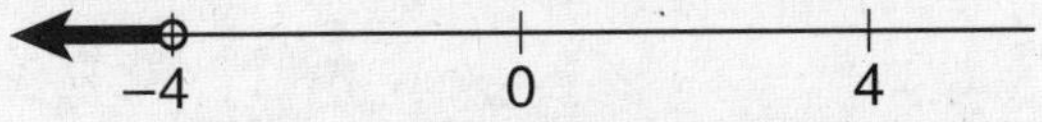

H. 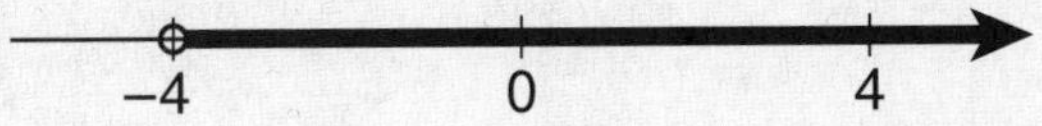

J. 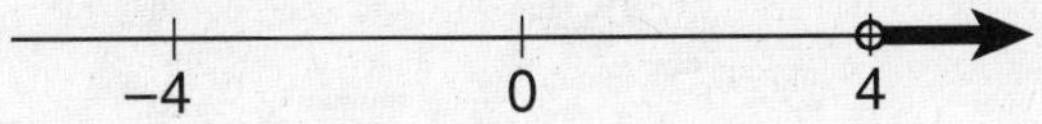

K. 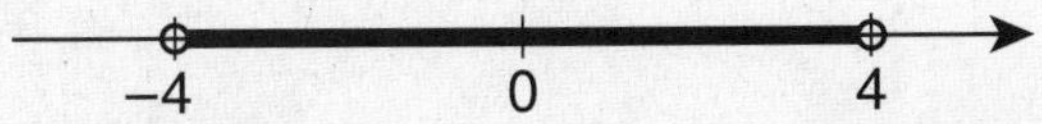

33. Which of the following is equivalent to $(a^2b)^x$?

A. $a^{2+x}b^x$

B. a^2b^x

C. $a^{2+x}b^{1+x}$

D. xa^2b

E. $a^{2x}b^x$

GO ON

ACT Practice Test 2 Section 1 *continued*

DO YOUR FIGURING HERE.

34. When the point $(4, -1)$ is reflected across the y-axis, what are the coordinates of its image?

F. $(-4, -1)$

G. $(-4, 1)$

H. $(-1, 4)$

J. $(4, -1)$

K. $(4, 1)$

35. If $(1, -2, 7)$ and $(2, 4, 2)$ are points in 3D-space, what is the approximate distance between the two points?

A. 4

B. 5

C. 6

D. 7

E. 8

36. If $a = 6$, then $\dfrac{\frac{a}{16}}{\frac{27}{a}} = ?$

F. $\frac{1}{72}$

G. $\frac{1}{12}$

H. $\frac{16}{27}$

J. $\frac{27}{36}$

K. $\frac{27}{16}$

37. How many common factors do 24 and 36 have?

A. 1

B. 2

C. 4

D. 6

E. 8

GO ON

Name ______________________ Date __________ Class __________

ACT Practice Test 2 Section 1 *continued*

DO YOUR FIGURING HERE.

38. If $x = 2\pi$, then $\cos(\pi + \frac{x}{4}) =$

F. -1

G. $-\frac{1}{2}$

H. 0

J. $\frac{1}{2}$

K. 1

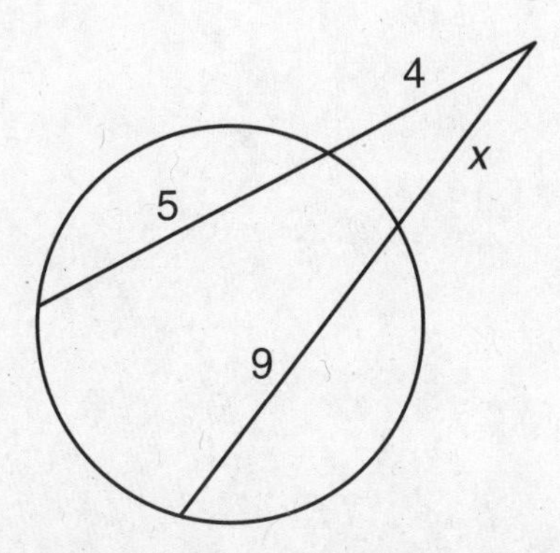

39. What is the value of x in the diagram above?

A. 0

B. 1

C. 2

D. 3

E. 4

40. If $x^2 + 5x - 14 > 0$, then which of the following can NOT equal x?

F. -21

G. -19

H. 1

J. 3

K. 19

GO ON

ACT Practice Test 2 Section 1 *continued*

DO YOUR FIGURING HERE.

41. Jean has 4 jackets, 5 hats, and 3 pairs of boots. If she wears one of each whenever it gets cold, how many different outfits can she wear in the cold weather?

A. 12

B. 20

C. 30

D. 40

E. 60

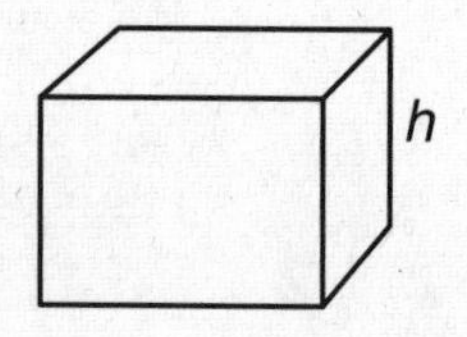

42. The solid above is a box and the length, width, and height are all integers. If the volume of the box is 32 cubic units, which of the following could NOT be the height?

F. 2

G. 4

H. 6

J. 8

K. 16

43. Joe has a set of data to study. He adds one more data value to the set and as a result, the range increased. Which of the following statements MUST be true?

A. The median also increased.

B. The mean also increased.

C. The mode also increased.

D. The mean did not change.

E. The mode did not change.

GO ON

Name ______________________________ Date ____________ Class ____________

ACT Practice Test 2 Section 1 *continued*

DO YOUR FIGURING HERE.

44. The minimum y-value of the graph of $y = -2\sin x + 2$ is

F. −4

G. −3

H. −2

J. −1

K. 0

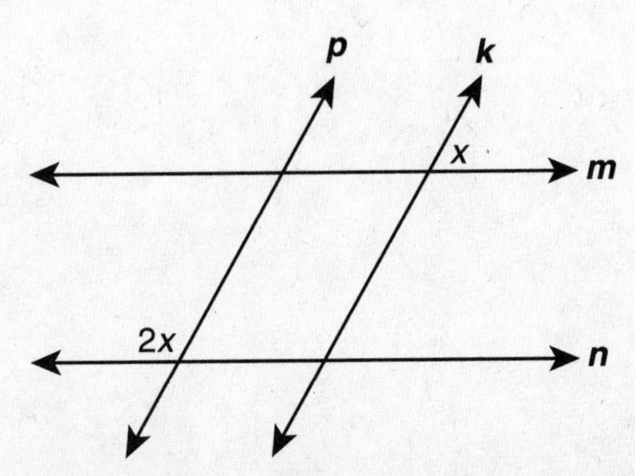

45. If lines m and n above are parallel and lines p and k are parallel, what is the value of x in degrees?

A. 30

B. 45

C. 60

D. 90

E. 120

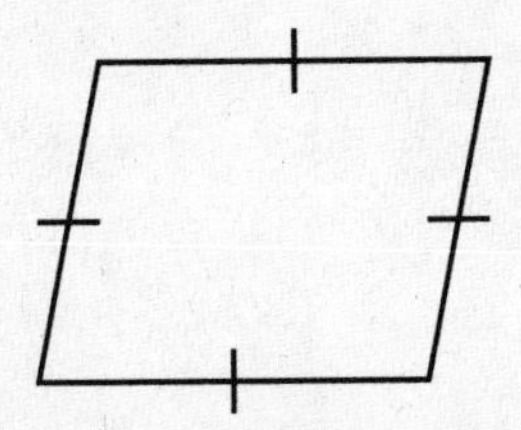

46. Which of the following statements about the polygon shown above is not ALWAYS true?

F. The figure is a square.

G. The figure is a rhombus.

H. The figure is a parallelogram.

J. The figure is a quadrilateral.

K. The figure is two-dimensional.

GO ON

Name ______________________ Date ____________ Class ____________

ACT Practice Test 2 Section 1 *continued*

DO YOUR FIGURING HERE.

47. Points *a*, *b*, and *c* all lie on a number line. If the distance from *b* to *c* is 3, the distance from *a* to *c* is 5, and *a* is -1, then *b* could NOT be which of the following?

A. -9

B. -4

C. -3

D. 1

E. 7

48. The value of Mrs. Jones' car decreases by 10% each year. If she paid $18,000 for her car in 2000, how much was the car worth in 2003?

F. $131

G. $180

H. $1312

J. $1800

K. $13,122

49. For the equation $ax^2 + bx + c = 0$, if $b^2 - 4ac = 0$, then the equation has

A. one real root

B. two real roots

C. one real root and one complex root

D. two complex roots

E. no roots

50. The measures of the angles of a triangle are in the ratio 6 : 9 : 10. What is the measure, in degrees, of the smallest angle?

F. 42.2

G. 43.2

H. 44.2

J. 45.2

K. 46.2

GO ON

Name ______________________ Date __________ Class __________

ACT Practice Test 2 Section 1 *continued*

Colors of Cars in the Parking Lot

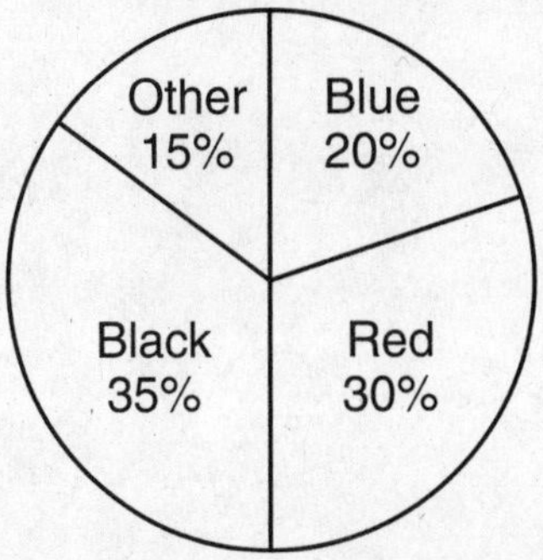

Other	15%
Blue	20%
Black	35%
Red	30%

DO YOUR FIGURING HERE.

51. There are 200 cars in a parking lot. According to the data above, which of the following could NOT be the number of yellow cars?

A. 0

B. 10

C. 20

D. 30

E. 40

52. If $x^3 + 3x^2 \leq 0$, then x can NOT be which of the following?

F. -10

G. -5

H. -3

J. -1

K. 0

53. If $x = -3$ and $y = |2x + 4| - |6 - 3x|$, then what does y equal?

A. -13

B. -5

C. 5

D. 13

E. 17

GO ON

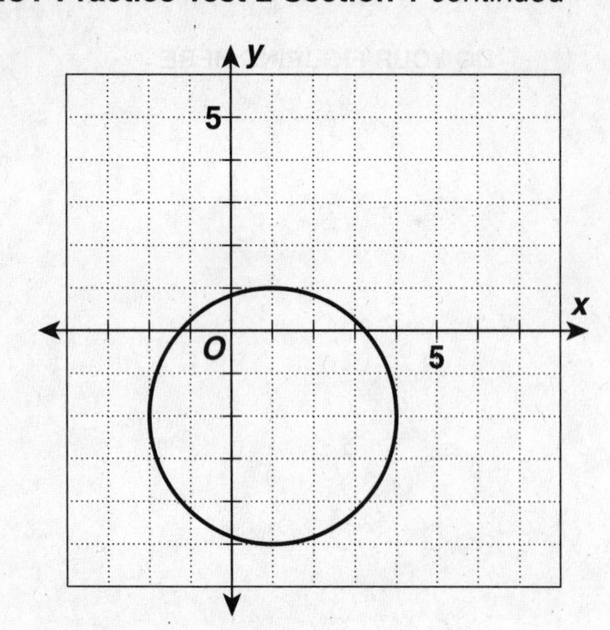

DO YOUR FIGURING HERE.

54. Which of the following could be the equation of the conic graphed?

F. $\frac{x^2}{3} + \frac{y^2}{3} = 1$

G. $\frac{x^2}{9} + \frac{y^2}{9} = 1$

H. $\frac{(x-1)^2}{9} + \frac{(y+2)^2}{9} = 1$

J. $\frac{(x-1)^2}{3} + \frac{(y+2)^2}{3} = 1$

K. $\frac{(x+1)^2}{9} + \frac{(y-2)^2}{9} = 1$

55. What is the product of the roots of the equation $(4x - 5)(25x + 2) = 0$?

A. -10

B. $-\frac{1}{10}$

C. $\frac{-3}{100}$

D. 10

E. $\frac{100}{3}$

GO ON

Name ______________________ Date __________ Class __________

DO YOUR FIGURING HERE.

56. Which of the following values of x satisfies the equation $\sin(2x - 1) + 5 = 5$?

F. 0

G. $\frac{1}{4}$

H. $\frac{1}{2}$

J. $\frac{\pi}{2}$

K. π

57. If $f(-2) = 5$, $f(0) = 1$, and $f(2) = 5$, which of the following is a valid representation of $f(x)$?

A. $f(x) = -x + 1$

B. $f(x) = x - 1$

C. $f(x) = x + 1$

D. $f(x) = x^2 - 1$

E. $f(x) = x^2 + 1$

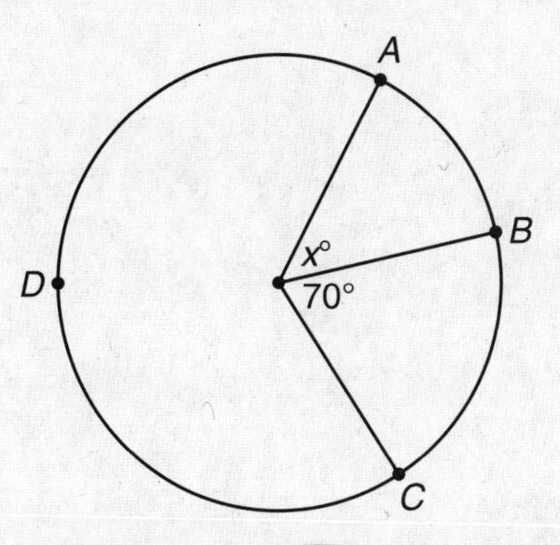

58. If the measure of $\overset{\frown}{ADC}$ is twice the measure of $\overset{\frown}{AC}$, what is the value of x in degrees?

F. 30

G. 40

H. 50

J. 60

K. 70

GO ON

Name ______________________ Date ____________ Class ____________

DO YOUR FIGURING HERE.

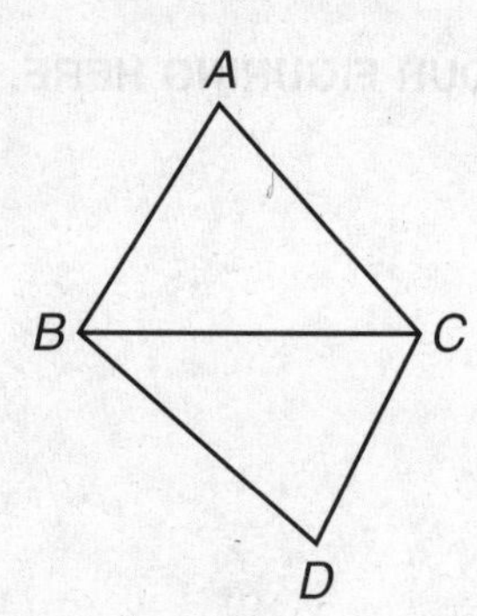

GIVEN: $AC = BD$ and $\angle ACB \cong \angle DBC$
PROVE: $\triangle ABC \cong \triangle DCB$

59. For the proof above, what would be the final justification?

A. SSA

B. SSS

C. ASA

D. SAS

E. AAA

60. Simplify $\dfrac{\frac{x}{3}}{\frac{x}{3} + \frac{3}{x}}$.

F. $\frac{1}{2}$

G. $\frac{1}{3}$

H. $\frac{x}{3}$

J. $\frac{3}{x+3}$

K. $\frac{x^2}{x^2+9}$

If you finish before time is called, you may check your work on this section only. Do not turn to any other section in the test.

Answer Key

PSAT Practice Test 1

Section 1

1. E
2. B
3. D
4. D
5. D
6. A
7. D
8. E
9. E
10. A
11. C
12. B
13. B
14. B
15. D
16. D
17. D
18. E
19. B
20. E

Section 2

1. B
2. C
3. A
4. A
5. A
6. B
7. C
8. C
9. 11
10. $\frac{2}{3}$
11. 4
12. 2
13. $\frac{2}{8}$ or 0.25
14. 56
15. 9
16. 8
17. 3.3
18. 1

PSAT Practice Test 2

Section 1

1. C
2. B
3. B
4. A
5. C
6. E
7. D
8. D
9. C
10. E
11. B
12. E
13. D
14. A
15. C
16. C
17. E
18. C
19. B
20. D

Section 2

1. C
2. E

3. E
4. A
5. A
6. E
7. D
8. C
9. 9
10. 6
11. 2
12. 85
13. $\frac{1}{25}$ or 0.04
14. 15
15. 25
16. 30
17. 17
18. 36

SAT Practice Test 1

Section 1

1. E
2. B
3. E
4. C
5. E
6. A
7. A
8. C
9. C
10. E
11. A
12. C
13. B
14. A
15. B
16. B
17. B
18. B
19. D
20. A

Section 2

1. E
2. B
3. D
4. B
5. E
6. C
7. D
8. E
9. 9
10. 30
11. 10
12. 1
13. $\frac{1}{8}$
14. 4
15. 18
16. 24.5
17. 7
18. 4

Section 3

1. E
2. B
3. E
4. A
5. E
6. D
7. C
8. E

9. B
10. C
11. E
12. D
13. B
14. A
15. B
16. D

SAT Practice Test 2

Section 1

1. D
2. E
3. D
4. B
5. B
6. C
7. D
8. D
9. A
10. E
11. C
12. A
13. B
14. C
15. B
16. D
17. A
18. B
19. E
20. E

Section 2

1. D
2. B
3. B
4. B
5. E
6. C
7. C
8. E
9. 12
10. 13
11. 1152
12. 11
13. $\frac{2}{3}$
14. 6
15. 6
16. 19
17. 200
18. 2.5

Section 3

1. B
2. D
3. B
4. A
5. A
6. E
7. C
8. D
9. C
10. A
11. C
12. C
13. D

14. B
15. C
16. C

SAT Subject Test Practice Test 1 Level IC

1. C
2. B
3. C
4. E
5. B
6. C
7. D
8. D
9. C
10. B
11. A
12. D
13. C
14. E
15. D
16. E
17. B
18. B
19. A
20. D
21. D
22. C
23. A
24. B
25. B
26. C
27. E
28. A
29. D
30. C
31. D
32. A
33. B
34. C
35. B
36. C
37. A
38. D
39. E
40. D
41. B
42. D
43. B
44. A
45. D
46. A
47. E
48. D
49. B
50. E

SAT Subject Test Practice Test 2 Level IC

1. B
2. D
3. D
4. E
5. B
6. C
7. A
8. D
9. A
10. A
11. E

12. D
13. C
14. D
15. D
16. A
17. B
18. D
19. C
20. E
21. A
22. C
23. B
24. B
25. B
26. D
27. D
28. E
29. D
30. B
31. A
32. C
33. E
34. E
35. B
36. C
37. B
38. E
39. D
40. C
41. C
42. A
43. B
44. D
45. C
46. D
47. B
48. C
49. B
50. A

SAT Subject Test Practice Test 1 Level IIC

1. A
2. B
3. E
4. B
5. D
6. E
7. D
8. A
9. D
10. A
11. B
12. D
13. B
14. C
15. C
16. C
17. B
18. A
19. C
20. A
21. C
22. D
23. E
24. B
25. E
26. C

27. B
28. C
29. A
30. D
31. B
32. E
33. E
34. C
35. B
36. C
37. B
38. D
39. B
40. C
41. E
42. A
43. E
44. B
45. B
46. D
47. D
48. A
49. D
50. C

SAT Subject Test Practice Test 2 Level IIC

1. B
2. B
3. E
4. E
5. C
6. A
7. B
8. C
9. D
10. C
11. D
12. E
13. E
14. B
15. D
16. C
17. B
18. C
19. D
20. A
21. C
22. B
23. C
24. C
25. D
26. A
27. E
28. E
29. A
30. E
31. C
32. E
33. E
34. E
35. C
36. B
37. A
38. A
39. D
40. E
41. D

42. B
43. B
44. A
45. C
46. C
47. D
48. C
49. E
50. A

ACT Practice Test 1

1. D
2. J
3. A
4. H
5. C
6. J
7. A
8. J
9. A
10. J
11. E
12. F
13. E
14. J
15. C
16. G
17. C
18. F
19. E
20. G
21. C
22. K
23. C
24. J
25. D
26. G
27. D
28. G
29. D
30. J
31. C
32. G
33. C
34. G
35. A
36. G
37. C
38. H
39. A
40. G
41. A
42. K
43. C
44. J
45. D
46. J
47. B
48. H
49. E
50. H
51. A
52. J
53. C
54. H
55. C
56. J

57. C
58. G
59. D
60. H

ACT Practice Test 2

1. D
2. F
3. D
4. F
5. D
6. F
7. E
8. K
9. C
10. H
11. C
12. J
13. B
14. J
15. A
16. H
17. C
18. H
19. E
20. J
21. A
22. H
23. C
24. G
25. D
26. H
27. B
28. G
29. A
30. K
31. B
32. H
33. E
34. F
35. E
36. G
37. D
38. H
39. D
40. H
41. E
42. H
43. E
44. K
45. C
46. F
47. B
48. K
49. A
50. G
51. E
52. J
53. A
54. H
55. B
56. H
57. E
58. H
59. D
60. K